Advances in Computer-Aided Technology II

Advances in Computer-Aided Technology II

Advances in Computer-Aided Technology II

Guest Editors

Martin Pollák
Marek Kočiško

MDPI

Basel • Beijing • Wuhan • Barcelona • Belgrade • Novi Sad • Cluj • Manchester

Guest Editors
Martin Pollák
Faculty of Manufacturing Technologies
Technical University of Kosice
Prešov
Slovakia

Marek Kočiško
Faculty of Manufacturing Technologies
Technical University of Kosice
Prešov
Slovakia

Editorial Office
MDPI AG
Grosspeteranlage 5
4052 Basel, Switzerland

This is a reprint of the Special Issue, published open access by the journal *Machines* (ISSN 2075-1702), freely accessible at: https://www.mdpi.com/journal/machines/special_issues/46J0IVH1GT.

For citation purposes, cite each article independently as indicated on the article page online and as indicated below:

Lastname, A.A.; Lastname, B.B. Article Title. *Journal Name* **Year**, *Volume Number*, Page Range.

ISBN 978-3-7258-6007-4 (Hbk)
ISBN 978-3-7258-6008-1 (PDF)
https://doi.org/10.3390/books978-3-7258-6008-1

Contents

Contents

About the Editors

Martin Pollák

Martin Pollák graduated in 2014 from the Faculty of Manufacturing Technologies of the Technical University of Kosice, and he completed his PhD. (2017) in the field of engineering technologies and materials. He currently works as an associate professor and is also the head of the Department of Computer-Aided Manufacturing Technologies at the Faculty of Manufacturing Technologies with a seat in Prešov, TUKE. His work focuses on CA systems, computer simulation, machine vision using UI, automation, and manufacturing technologies, with a focus on the use of additive technology in robotics. His strengths include knowledge of working in CAD/CAM/CAE systems and simulation systems for the design of robotic workplaces, programming of industrial, collaborative, and educational robots, diagnostics of production equipment, and work with programming languages. He has extensive experience in solving scientific research and design tasks, which he actively applies in the process of design and programming. He is a reviewer of many journals registered in Current Contents, Web of Science, and Scopus databases.

Marek Kočiško

Marek Kočiško graduated in 2001 from the Faculty of Manufacturing Technologies of the Technical University of Košice and he completed his PhD. (2004) in the field of engineering technologies and materials. He currently works as a full professor and is also the vice-dean for development and external relations at the Faculty of Manufacturing Technologies with a seat in Prešov, TUKE. His work focuses on computer-aided production processes, CAD/CAM/CAE systems, simulation of production systems, computer networks, and virtual (augmented) reality, as well as reverse and simultaneous engineering. His strengths include knowledge of PC hardware, computer networks, notification technology, excellent knowledge of operating systems, office systems, computer simulation systems, and CAD/CAM/CAE systems (Creo, Unigraphics NX, Solid Edge, etc.). He is an expert member of many national and international associations and also a reviewer of many journals registered in Current Contents, Web of Science, and Scopus databases.

machines

Editorial

Editorial for Special Issue "Advances in Computer-Aided Technology II"

Martin Pollák * and Marek Kočiško

Faculty of Manufacturing Technologies, Technical University of Kosice, Bayerova 1, 080 01 Presov, Slovakia; marek.kocisko@tuke.sk
* Correspondence: martin.pollak@tuke.sk

The selection of the topic for this Special Issue, Advances in Computer-Aided Technology II, under the section "Mechatronic and Intelligent Machines", has garnered positive feedback from the scientific community. Computer-aided technologies (CAx) refers to the integration of computer technology in the design, analysis, and production of products, and noteworthy advancements have occurred in this field due to the swift evolution of science and technology. Advanced CAx tools incorporate various elements of product lifecycle management (PLM), such as design, finite element analysis (FEA), manufacturing processes, and production planning. With the shift toward the Industry 4.0 paradigm, the concept of a digital twin has become increasingly important, necessitating the adaptation of current CAx systems. This Special Issue features eight chosen papers that address industrial applications across significant topics within the scientific community, including emerging trends in CAx systems, additive manufacturing, neural networks and IoT in manufacturing, production system simulations, robotics, advanced FEA systems, material engineering, digitization, and virtual and augmented reality.

A study by Ling et al. presents an enhanced extreme learning machine (ELM) neural network to connect robotic spraying processes with quality outcomes, addressing the complexities of manual spraying that relies on human expertise. An optimization using a K-means improved predator optimization algorithm (KHPO) significantly improves the model's prediction accuracy, achieving up to a 61.95% reduction in errors for average film thickness compared to traditional ELM models. This KHPO-ELM neural network effectively predicts multi-spraying parameters, thereby facilitating improved automated spraying quality [1]. ISO 9283 serves as an important standard for evaluating robot performance. The paper of authors Pollák et al. focuses on verifying the repeatability of the Panasonic TM-2000 welding robot in a manufacturing setting. The study utilizes simulation software RoboDK to create a complete welding station, involving the design of a measuring device and the determination of measurement protocols. The results, based on the measurements taken from multiple points along the X and Y axes, were compared to manufacturer specifications, illustrating the robot's positional repeatability with graphical representations of its deviations [2].

Additive manufacturing is increasingly used across various fields, but its components exhibit orthotropic behavior, making mechanical properties difficult to predict due to various influencing factors. The study by Grozav et al. developed a predictive model using an artificial neural network trained on empirical tensile strength data from FDM-printed PLA test parts, achieving an accuracy of approximately 93%. The model successfully correlates predicted mechanical characteristics with real-world testing, allowing for an accurate FEA of complex geometries by defining custom orthotropic material profiles [3].

Furthermore, the article by Novikov et al. introduces a new methodology for evaluating modern visualization tools in the pre-production phase of manufacturing processes, validated through the successful implementation of ten real projects. The methodology uses a modified Z-score to categorize processes and facilitates decision-making for manufacturing companies in deploying computer-aided instruction tools such as tutorials utilizing video, augmented reality, and virtual reality. The methodology confirms a dependency between the chosen visualization tools and the nature of production processes, enhancing the rationalization of pre-production activities [4].

Currently, there is a growing emphasis on reducing the carbon footprint of production processes, and natural fiber-reinforced plastics (NFRP) have potential applications due to their unique properties. Conventional machining of NFRP faces challenges because of their non-homogenous structures, but abrasive water jet (AWJ) machining offers a solution. The article by Mitalova et al. evaluates the surface topography of an NFRP composite after AWJ cutting, demonstrating promising results for the application of this technology in processing NFRP materials [5]. The paper by Popan et al. addresses the challenge of enhancing the accuracy of abrasive waterjet cutting (AWJC) for carbon-fiber-reinforced polymers (CFRP), focusing on minimizing shape errors, particularly at the cut-in and cut-out points. The study investigates the formation mechanisms of these errors related to lead-in/lead-out strategies and process parameters, proposing a correction method with recommendations for the selection of an optimal strategy. A mathematical model to determine the depth of shape errors was developed and validated through industrial applications [6]. The paper by authors Vandzura et al. examines the direct metal laser sintering (DMLS) process, which is widely utilized in additive manufacturing across various industries, particularly for the production of prototypes and small series using metal powders like aluminum alloys and maraging steels. The study focuses on optimizing heat treatment processes for maraging steel samples printed via DMLS, experimenting with reduced maintenance times at elevated temperatures to achieve desired hardness levels. The results indicate that maintaining a temperature of 590 °C for just 1 h, followed by air cooling, achieves a hardness of around 50 HRC while also yielding a fine and uniform microstructure [7].

The paper of Rojas-Sola et al. analyses the design of the single-cylinder steam engine of the Grasshopper beam created by Henry Muncaster in 1912 and later published in the Model Engineer journal in 1957. A computer-aided engineering (CAE) study was conducted using Autodesk Inventor Nastran to perform a linear static analysis of the 3D CAD model under real operating conditions, focusing on material resistance criteria. The analysis identified the maximum admissible steam pressure during operation as 0.15 MPa, while evaluating von Mises stresses, displacements, and safety coefficients under unfavorable situations [8].

The editorial team expresses gratitude to all of the authors of the published articles, as well as those who showed interest in this Special Issue. Special thanks are also extended to the reviewers for their thorough assessment of the submitted manuscripts, whose expertise contributed valuable insights and recommendations for enhancing the authors' research findings.

Conflicts of Interest: The authors declare no conflicts of interest.

References

1. Ling, L.; Zhang, X.; Hu, X.; Fu, Y.; Yang, D.; Liang, E.; Chen, Y. Research on Spraying Quality Prediction Algorithm for Automated Robot Spraying Based on KHPO-ELM Neural Network. *Machines* **2024**, *12*, 100. [CrossRef]
2. Pollák, M.; Goryl, K. Simulation Design and Measurement of Welding Robot Repeatability Utilizing the Contact Measurement Method. *Machines* **2023**, *11*, 734. [CrossRef]
3. Grozav, S.D.; Sterca, A.D.; Kočiško, M.; Pollák, M.; Ceclan, V. Artificial Neural Network-Based Predictive Model for Finite Element Analysis of Additive-Manufactured Components. *Machines* **2023**, *11*, 547. [CrossRef]
4. Novikov, K.; Hořejší, P.; Kubr, J.; Dvořák, M.; Bednář, M.; Krákora, D.; Krňoul, M.; Šimon, M. Methodology for Rationalization of Pre-Production Processes Using Virtual Reality Based Manufacturing Instructions. *Machines* **2024**, *12*, 2. [CrossRef]
5. Mitaľová, Z.; Klichová, D.; Botko, F.; Litecká, J.; Vandžura, R.; Mitaľ, D. Green Machining of NFRP Material. *Machines* **2023**, *11*, 692. [CrossRef]
6. Popan, I.A.; Cosma, C.; Popan, A.I.; Panc, N.; Filip, D.; Balc, N. Correction of Shape Error at Cut-In and Cut-Out Points in Abrasive Waterjet Cutting of Carbon Fiber Reinforced Polymer (CFRP). *Machines* **2023**, *11*, 800. [CrossRef]

7. Vandzura, R.; Simkulet, V.; Gelatko, M.; Hatala, M.; Mitalova, Z. Effect of Hardening Temperature on Maraging Steel Samples Prepared by Direct Metal Laser Sintering Process. *Machines* **2023**, *11*, 351. [CrossRef]
8. Rojas-Sola, J.I.; Gutiérrez-Antúnz, J.F. Analysis of the Design of the Single-Cylinder Steam Engine of the Grasshopper Beam by Henry Muncaster. *Machines* **2023**, *11*, 703. [CrossRef]

machines

MDPI

Article

Research on Spraying Quality Prediction Algorithm for Automated Robot Spraying Based on KHPO-ELM Neural Network

Le Ling [1,†], Xuejian Zhang [2,3,†], Xiaobing Hu [2,3,*], Yucong Fu [2,3], Dongming Yang [2,3], Enpei Liang [1] and Yi Chen [1]

1 Dongfang Electric Group Science and Technology Research Institute Company Limited, No. 18 Xixin Avenue, Gaoxin West District, Chengdu 611731, China; lingle@hotmail.com (L.L.); liangep0670@dongfang.com (E.L.); chenyi02@dongfang.com (Y.C.)

2 School of Mechanical Engineering, Sichuan University, No. 24, South Section 1, First Ring Road, Chengdu 610065, China; advtechlab@stu.scu.edu.cn (X.Z.); fuyucong@stu.scu.edu.cn (Y.F.); 2021223025108@stu.scu.edu.cn (D.Y.)

3 Institute of Industrial Technology, Sichuan University, Changjiang North Road, Yibin 644005, China

* Correspondence: huxb@scu.edu.cn

† These authors contributed equally to this work.

Abstract: In the intelligent transformation of spraying operations, the investigation into the robotic spraying process holds significant importance. The spraying process, however, falls within the realm of experience-driven technology, characterized by high complexity, diverse parameters, and coupling effects. Moreover, the quality of manual spraying processes relies entirely on manual experience. Thus, the crux of the intelligent transformation of spraying robots lies in establishing a mapping model between the spraying process and the resultant spraying quality. To address the challenge of intelligently transforming empirical spraying processes and achieving the mapping from the spraying process to spraying quality, an algorithm employing an enhanced extreme learning machine-based neural network is proposed for predicting spraying process parameters with respect to the evaluation index of spraying quality. In this approach, an algorithmic model based on the Extreme Learning Machine (ELM) neural network is initially constructed utilizing five spraying process parameters: spraying speed, spraying height, spraying width pressure, atomization pressure, and oil spraying pressure. Two spraying quality evaluation indexes, namely average film thickness at the center point and surface roughness, are also incorporated. Subsequently, the prediction neural network is optimized using the K-means improved predator optimization algorithm (KHPO) to enhance the model's prediction accuracy. This optimization step aims to improve the efficiency of the model in predicting spraying quality based on the specified process parameters. Finally, data collection and model validation for the spraying quality prediction algorithm are conducted using a designed robotic automated waterborne paint spraying experimental system. The experimental results demonstrate a significant reduction in the prediction error of the KHPO-ELM neural network model for the average film thickness center point, showcasing a decrease of 61.95% in comparison to the traditional ELM neural network and 50.81% in comparison to the BP neural network. Likewise, the improved neural network model yields a 2.31% decrease in surface roughness prediction error compared to the traditional ELM neural network and a substantial 54.0% reduction compared to the BP neural network. Consequently, the KHPO-ELM neural network, incorporating the prediction algorithm, effectively facilitates the prediction of multi-spraying process parameters for the center point of average film thickness and surface roughness in automated robot spraying. Notably, the prediction algorithm exhibits a commendable level of accuracy in these predictions.

Keywords: robotic spraying process; spraying process parameters; spraying quality evaluation index; KHPO; ELM

1. Introduction

Spraying is a method of dispersing paint into uniform and fine droplets and applying it to the surface of an object by means of a spray gun, using air pressure, which serves to cover and protect [1]. Spraying is a very important process in the processing of many products. It can provide anti-corrosion protection, insulation protection, wear protection and other auxiliary protection measures for mechanical parts in order to improve operating efficiency and service life in various types of harsh working environments [2]. The spraying process is an experience-driven technology [3]; the operation needs to rely on the operator's production experience to adjust the spraying process parameters. Therefore, a vast majority of factories still use manual spraying methods, which not only leads to huge labor costs, but also to low efficiency and unstable quality. In addition, harsh working environments and toxic and harmful substances in paints pose an occupational health hazard to the operators [4].

To address the challenges of inefficiency, unstable quality, and health hazards associated with manual spraying, technological upgrading of the spraying process is imperative. The transformation involves elevating the manual spraying process to an automated spraying production line with robotic operation [5]. This advanced production line seamlessly integrates industrial robotics technology with emerging Internet technologies such as neural networks and fuzzy control. It facilitates the learning and reuse of empirical manual processes, retaining the guiding role of manual experience while capitalizing on the unparalleled advantages of manual work in terms of iteration and generalization [6]. In comparison to manual spraying, robotic spraying boasts superior advantages, including heightened productivity, reduced production costs, enhanced spraying quality and quantity, optimal utilization of paint, and minimized health risks.

Despite these advantages, the design of the robotic spraying process is highly intricate owing to the numerous spraying process parameters involved and their strong coupling relationships [7]. For spraying robots, the core technical indicator is the quality of spraying. Balancing improved productivity with the achievement of high-quality spraying operations is crucial in the realm of robot spraying. Key evaluation indices for spraying quality primarily encompass the paint layer thickness and surface roughness of the sprayed components [8]. The current focal point of research in the robotic spraying process direction lies in establishing the role of the spraying parameter model and integrating spraying process modeling with spraying trajectory planning. Wu, HJ et al. introduced a method capable of predicting spraying thickness on complex part surfaces affected by shadowing effects. The method involves establishing a three-dimensional geometric model of the spraying contour based on Gaussian distribution, integrating robot trajectory and machining parameters. This approach not only creates a 3D geometric model of the spraying profile but also simulates spraying deposition in robot offline programming software. It concludes by visualizing the spraying morphology and predicted spraying thickness in a graphical virtual environment [9]. Zhang, YJ et al. proposed a predictive analytical model for the thickness of a spherical coated surface. Through numerical simulation and experimental testing, they demonstrated the model's effectiveness in predicting spraying thickness and uniformity with acceptable tolerance, particularly for spherical surfaces [10]. Gleeson, D et al. presented a paint spraying optimization method to refine the generated initial trajectory and minimize paint thickness deviation from the target thickness [11]. Gadow, R et al. introduced an advanced computer-aided robotic path planning method for thermal spraying. This method integrates optimization of generated trajectories and velocity profiles into coupled Computational Fluid Dynamics (CFD) and Finite Element Method (FEM) models, analyzing their effects on heat and mass transfer during spraying deposition [12]. Potkonjak, V et al. optimized the gun trajectory based on ergonomics and developed a spraying model for arbitrary gun position and attitude. The optimization resulted in reduced motor load, energy consumption, and need to control jitter [13]. Arikan, MAS et al. developed a methodology and computer program for modeling the painting process, simulating robotic painting, offline programming of industrial robots, and measuring film thickness for curved surface painting. This comprehensive approach enables the simulation and analysis of spraying thickness [14].

While numerous relevant studies exist, several challenges persist. Firstly, the mathematical modeling of film thickness calculation exhibits a substantial gap between theoretical calculations and actual production and processing on-site, resulting in a lack of robustness in the film thickness calculation model. Secondly, the majority of studies on spraying distance, spray pressure, and other spraying parameters rely on qualitative research for simulation models. Due to the strong coupling effect among these parameters, the qualitative conclusions derived from such studies often diverge from actual processing, resulting in low feasibility in real-world applications. Consequently, utilizing field process parameter data and establishing a neural network-based prediction algorithm model incorporating multiple spraying process parameters emerges as the optimal solution to address the complexities of the field environment, offering high robustness, adaptability, and rapid iteration capabilities. This approach using neural networks for predicting and analyzing the quality of the spraying process presents advantages that the simulation model research method lacks [15].

Firstly, neural networks can model complex relationships owing to their substantial nonlinear fitting ability, allowing them to capture intricate relationships between input and output parameters. This capability enables the establishment of mapping relationships that explicit mathematical models cannot construct [16]. Secondly, neural networks exhibit strong generalization ability, providing reliable prediction results even in the face of unknown working conditions and parameter combinations by generalizing from training samples. Thirdly, neural networks allow flexible iteration, permitting the optimization of the combination of spraying process parameters through the correction of model training samples.

To address the aforementioned challenges, a prediction algorithm for spraying center film thickness and surface roughness based on multiple spraying process parameters and an improved Extreme Learning Machine neural network is proposed. This algorithm leverages neural networks, starting from the actual spraying processing site, to predict and evaluate the impact of spraying process parameters. The algorithm begins by studying the relationship between various spraying parameters and commonly used spraying quality evaluation indexes, selecting input and output parameters for the prediction model. Subsequently, the algorithm introduces the Extreme Learning Machine (ELM) neural network prediction algorithm model based on multiple spraying process parameters to map the spraying parameters to the spraying quality evaluation indexes. Finally, the prediction model for spraying quality evaluation indexes is optimized using the hyperparameter optimization (HPO) algorithm and the K-means clustering optimization algorithm. This optimization aims to enhance the convergence speed and global optimization ability of the prediction model. The proposed method enables the evaluation of spraying quality before robotic spraying operations, streamlining the optimization of the spraying process, reducing iterative costs associated with spraying operations, and ultimately enhancing the efficiency of process optimization.

2. Spraying Process

2.1. Spraying Process Parameters

This paper exemplifies the automated spraying of waterborne paints on flat parts to investigate the impact of spraying process parameters on spraying quality. Spraying quality is primarily influenced by spraying powder, the spraying process, and spraying equipment. Among these factors, the role of the spraying process is the most intricate and crucial [17]. In conjunction with the trajectory planning requirements of the spraying robot and the influence of process parameters, this paper consolidates the experimental scenarios established, focusing on the more pronounced effects of certain process parameters on spraying quality. These parameters include spraying speed, spraying height, spraying width pressure, atomization pressure, and oil spraying pressure.

Spraying speed refers to the velocity of paint particles ejected from the gun relative to the surface of the workpiece being processed. This parameter directly influences the thickness and uniformity of the paint layer. Excessive spraying speed leads to a too-thin layer of paint, which fails to adhere to the workpiece surface, while insufficient speed results in an overly thick layer, possibly leading to solidification or undesired flow.

Spraying height denotes the relative distance between the tip of the spray gun and the workpiece surface. Excessive or insufficient spraying height can cause an uneven paint layer. Spray width pressure refers to the gas pressure needed to create a fan-shaped pattern at the tip of the gun. This pressure significantly influences the distribution of spray particles and the stability of the spray pattern. Proper spray width pressure ensures uniformity in spraying. Atomizing pressure refers to the gas pressure at the tip of the gun responsible for atomizing paint into fine particles. This pressure greatly impacts the size and uniformity of sprayed particles. Adjusting atomizing pressure allows for further refinement of paint particles, contributing to a uniform coating. Oil spraying pressure represents the force propelling the liquid paint. The adjustment of oil spraying pressure affects the flow of paint and the size of the sprayed particles.

A robust observation reveals a pronounced coupling among the effects of spraying process parameters. The influence of a solitary parameter extends beyond its impact on coating quality, extending to the "gain" or "offset" of other parameters.

For instance, heightened spraying speeds may yield excessively thin coatings, whereas slower speeds may result in coatings of excessive thickness. Simultaneously, the spray height significantly influences coating thickness. Consequently, by concurrently adjusting spray speed and spray distance to strike a balance between speed and height, it becomes feasible to elevate the spray speed while concurrently achieving a coating thickness and uniformity aligning with process requirements. Furthermore, a coupling exists between spraying speed and spray width pressure. Elevated spraying speeds might necessitate higher spray width pressure, yet an excessively high pressure could yield particles of diminished size, adversely affecting coating uniformity. Hence, a careful equilibrium between spray speed and pressure is imperative to ensure that accelerated spray speeds do not compromise particle size, preserving coating uniformity. Additionally, a coupling relationship extends to spraying pressure parameters. Higher atomizing pressure may produce excessively small particles, while lower spraying pressure may result in uneven spray distribution. Consequently, consideration of spray pressure becomes integral when adjusting atomization pressure. Strategic adjustments guarantee a sufficiently high atomizing pressure, while maintaining the spray pressure within an optimal range to achieve uniform coating.

The intricate coupling among spraying parameters profoundly influences spraying quality. Consequently, in the analysis of spraying quality, it is imperative not only to scrutinize the impact of individual parameters but also to factor in the intricate interplay among multiple spraying process parameters.

2.2. Spraying Quality Evaluation Indexes

The quality evaluation of the paint film completed by the spraying robot involves the analysis and judgment of various characteristics, such as performance testing and characterization, to ensure compliance with processing requirements [18]. The inspection of coating quality primarily includes the detection of indicators like film thickness, adhesion, surface roughness, appearance, hardness, and corrosion resistance.

With consideration given to the designated experimental scenarios and the significance of quality indicators, the selection of evaluation indices in this paper focuses on the average film thickness and surface roughness at the center point. Due to the normal distribution of coating thickness in the robotic spraying process, the average film thickness at the center point is commonly utilized for evaluating the film thickness index. The center point's uniform film thickness, defined as the average of film thickness values measured at various locations on the sprayed surface [19], directly signifies the uniformity of the paint film's surface area, a crucial aspect for coating quality.

Surface roughness, typically described in terms of surface parameters, characterizes the degree of microscopic unevenness on the surface of the object after spraying completion. The impact of surface roughness is direct on the adhesion, durability, and other properties of the paint layer.

Ensuring precision in high-precision mechanical products necessitates the meticulous control of the average film thickness and surface roughness at the coating center. This imperative is notably pertinent in the case of aero-engine blades and spacecraft housings, where the coating's height and surface roughness wield a considerable influence over aerodynamic performance, fuel efficiency, and structural durability during flight. By regulating the coating height and roughness, the hydrodynamic properties of the surface can be manipulated, thereby mitigating wind resistance and enhancing flight stability. Moreover, maintaining the coating within the ideal operational parameters effectively diminishes flight drag and surface friction, consequently augmenting fuel efficiency. In addition, adept control over coating height and roughness enables effective management of surface temperature and corrosion resistance. This adjustment, in turn, facilitates an extended service life and heightened reliability of the aircraft under challenging conditions, including high temperatures.

3. KHPO-ELM

3.1. ELM

The Extreme Learning Machine (ELM), introduced in 2006, is a type of single hidden layer feedforward neural network. It requires the initiation of hidden layer weights only once during training [20]. The ELM eliminates the need for backpropagation. Its accuracy depends on the number of models learned in training, resulting in quicker training speeds, simpler parameter adjustment, weight updates, and improved generalization ability.

The structure of the ELM network is depicted in Figure 1. In the training process of the ELM model, input layer weights and biases are randomly set, and output layer weights are calculated using generalized inverse matrix theory. Subsequently, data prediction is carried out by outputting information based on the obtained weights. The construction of the ELM neural network comprises four distinct stages.

1. According to the prediction objective, training sets X_i and t_i are set up with n samples.

$$X_i = [x_{i1}, x_{i2}, x_{i3}, \ldots, x_{in}]^T \in R^n, \tag{1}$$

$$t_i = [t_{i1}, t_{i2}, t_{i3}, \ldots, t_{im}]^T \in R^m, \tag{2}$$

   where x_i denotes the i_{th} data of the input matrix, and t_i denotes the labeling corresponding to the i_{th} data.

2. The hidden layer output o_j with L hidden layer nodes is computed.

$$o_j = \sum_{i}^{L} \beta_i g(W_i \cdot X_j + b_i), j = 1, 2, \ldots, N, \tag{3}$$

$$W_i = [w_{i,1}, w_{i,2}, w_{i,3}, \ldots, w_{i,n}]^T, \tag{4}$$

   where $g(x)$ is the activation function, W_i is the input weight, β_i is the output weight, b_i is the bias of the i_{th} hidden layer unit, and $W_i \cdot X_j$ is the inner product of W_i with X_j.

3. The output error is minimized by the existence of W_i, β_i, b_i, leading to the following Equation (5) holding true.

$$\sum_{j=1}^{N} \|o_j - t_j\| = 0, \tag{5}$$

$$t_j = \sum_{j=1}^{L} \beta_i g(W_i \cdot X_j + b_i), j = 1, 2, \ldots, N, \quad (6)$$

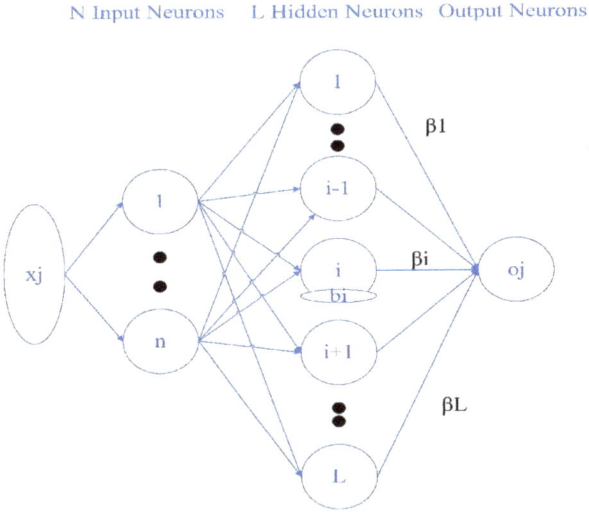

Figure 1. Structural diagram of the ELM neural network.

Leading to the matrix expression as:

$$H\beta = T, \quad (7)$$

$$H(W_1, W_2, \ldots, W_L, b_1, b_2, \ldots, b_L, X_1, X_2, \ldots, X_L) = \begin{bmatrix} g(W_1 \cdot X_1 + b_1) & g(W_L \cdot X_1 + b_L) \\ g(W_1 \cdot X_N + b_1) & g(W_L \cdot X_N + b_L) \end{bmatrix}_{N \times L} \quad (8)$$

$$\beta = \begin{bmatrix} \beta_1^T \\ \vdots \\ \beta_L^T \end{bmatrix}_{L \times m} \quad T = \begin{bmatrix} T_1^T \\ \vdots \\ T_N^T \end{bmatrix}_{N \times m}, \quad (9)$$

where H is the implicit layer node output, β is the output weight and T is the desired output.

The minimization loss function is E:

$$E = \sum_{j=1}^{N} \left(\sum_{j=1}^{L} \beta_i g(W_i \cdot X_j + b_i) - t_j \right)^2, \quad (10)$$

Through training, the optimal combination of $\hat{W}_i$, $\hat{\beta}_i$ and $\hat{b}_i$ parameters is obtained to maximize the model prediction accuracy.

3.2. KHPO

Due to the random generation of initialization weights and bias selection in the ELM neural network model, the accuracy is significantly impacted by different initial parameter settings. Additionally, the manual setting of hidden layer nodes in the model, along with the use of different node types, also plays a crucial role in performance. Consequently, employing optimization algorithms for global optimization and the automatic configuration of initial values and the number of hidden layer nodes in the ELM model holds significant importance in enhancing accuracy and convergence speed.

In this study, the ELM neural network model is optimized using a predator optimization algorithm. Proposed in 2022, the predator optimization algorithm achieves global optimization and rapid convergence by simulating the hunting process of animals [21]. The algorithm mimics the hunting process by continuously iterating the population search agent positions through the position update of the hunter and prey. It calculates the average position of the population, determines the distance between the search agent and the average agent using Euclidean distance, eliminates the maximum distance, and ultimately identifies the optimal position for optimization search.

The HPO algorithm flow is described below.

1. Initialization.

A random initialization community x_i is generated within the search range.

$$x_i = rand(1,d) * (ub - lb) + lb, \tag{11}$$

$$lb = [lb_1, lb_2, \cdots, lb_d], \tag{12}$$

$$ub = [ub_1, ub_2, \cdots, ub_d], \tag{13}$$

where x_i is the location of the search agent, lb is the lower bound of the variable, ub is the upper bound of the variable.

2. Hunter Location Updates;

$$x_{i,j}(t+1) = x_{i.j}(t) + 0.5\left[\left(2CZP_{pos(j)} - x_{i.j}(t)\right) + (2(1-C)Z\mu(j) - x_{i.j}(t))\right] \tag{14}$$

$$P = \vec{R_1} < C, \tag{15}$$

$$IDX = (P == 0), \tag{16}$$

$$Z = R_2 \otimes IDX + \vec{R_3} \otimes (\sim IDX), \tag{17}$$

$$C = 1 - it\left(\frac{0.98}{MaxIt}\right), \tag{18}$$

where $x_{i.j}(t)$ is the current position of the hunter, $x_{i,j}(t+1)$ is the position of the hunter in the next cycle, C is the balance parameter, $P_{pos(j)}$ is the position of the prey, μ is the average value of the position, i denotes the number of cycles, j denotes the number of iterations, $\vec{R_1}$ and $\vec{R_3}$ are the random vectors in [0, 1], P is the index value of $\vec{R_1} < C$, R_2 is the random value in [0, 1], IDX is the index value when the constant P is equal to 0, it is the number of current iterations, and $MaxIt$ is the maximum number of iterations.

3. Prey location update;

Prey position p_{pos} is calculated:

$$\vec{p_{pos}} = \vec{x_i} \in sortedD_{euc}(kbest)p_{pos}, \tag{19}$$

$$\mu = \frac{1}{n}\sum_{i=1}^{n}\vec{x_i}, \tag{20}$$

$$D_{euc(i)} = \left(\sum_{j=1}^{d}(x_{i.j} - \mu_j)^2\right)^{\frac{1}{2}}, \tag{21}$$

$$kbest = round(C \times N), \tag{22}$$

where $\vec{p_{pos}}$ denotes the search agent with the largest distance from the mean. N is the number of search agents. After the location of the prey is determined, it is captured by the hunter through the decreasing mechanism $kbest$.

The prey location update formula is:

$$x_{i.j}(t+1) = T_{pos(j)} + CZ\cos(2\pi R_4) \times \left(T_{pos(j)} - x_{i,j}(t)\right) \quad (23)$$

where $x_{i,j}(t)$ is the current position of the prey, $x_{i.j}(t+1)$ is the position of the next iteration of the prey, $T_{pos(j)}$ is the global optimal position, and R_4 is a random number within $[-1, 1]$.

In the HPO optimization algorithm, when the initial search range is extensive, there is a noticeable reduction in both convergence speed and optimization finding ability. Therefore, this paper adopts the K-means algorithm for a secondary optimization of the initialization process in the HPO algorithm. The K-means algorithm is employed to partition the search space into different clusters, enhancing search efficiency and diminishing the time required for convergence [22].

Using the K-means algorithm, the samples within the initial interval of values for LEM parameters (HPO search space) are organized into k classes. The N data within the specified interval are distributed into k classes, ensuring that each sample has the minimum distance from the center of its respective class. This process can be considered as a potential grouping of hunters or prey during the initialization of the HPO, with each cluster having its own initial solution.

4. Method of Prediction

The evaluation index of spraying quality, the average film thickness, and surface roughness at the center point are subject to the combined effect of spraying speed, spraying distance, spraying width pressure, atomization pressure, and oil spraying pressure, which are expressed by the Equation (24):

$$(H_{avg}, R_a) \in D = \{D | D(x) = f(V, D, P_{spray}, P_{atomization}, P_{oil})\} \quad (24)$$

where H_{avg} represents the average film thickness at the center point; R_a represents the surface roughness; S is the distribution model obeyed by the quality parameters with regard to spraying; S has a close relationship with the spraying process parameters; V represents the spraying speed; S is the spraying distance; P_{spray} is the spraying width pressure; $P_{atomization}$ is the atomization pressure; P_{oil} is the oil spraying pressure. A detailed description of the symbols for each coating process parameter is provided in Appendix A.

Due to the existence of coupling between the spraying process parameters and the complexity of solving the explicit mathematical formulas, the mapping model of the spraying process parameters V, S, P_{spray}, $P_{atomization}$, P_{oil} to H_{avg} and R_a can be established by pre-experimental data samples of the multi-parameter role of the relationship between the data fitting and prediction of the results.

Founded on the advanced KHPO-ELM neural network framework, this paper introduces a predictive model employing a "5-input-2-output" neural network architecture tailored for forecasting multiple quality assessment metrics associated with diverse spraying parameters, as delineated in Figure 2. The model facilitates the prognosis of the average film thickness and surface roughness at the central point, predicated upon variables such as spraying speed, spraying distance, spraying width pressure, atomizing pressure, and spraying pressure.

Furthermore, with regard to the model, this study employs the HPO optimization algorithm to enhance the ELM neural network. Within this context, the global optimization capability and convergence speed of the ELM neural network are heightened through the optimization of input weights, output weights, and biases of hidden layer units. Additionally, the initialization process of the HPO algorithm undergoes quadratic optimization utilizing the K-means algorithm, thereby enhancing the search efficiency and iteration speed of the initial population in the HPO algorithm.

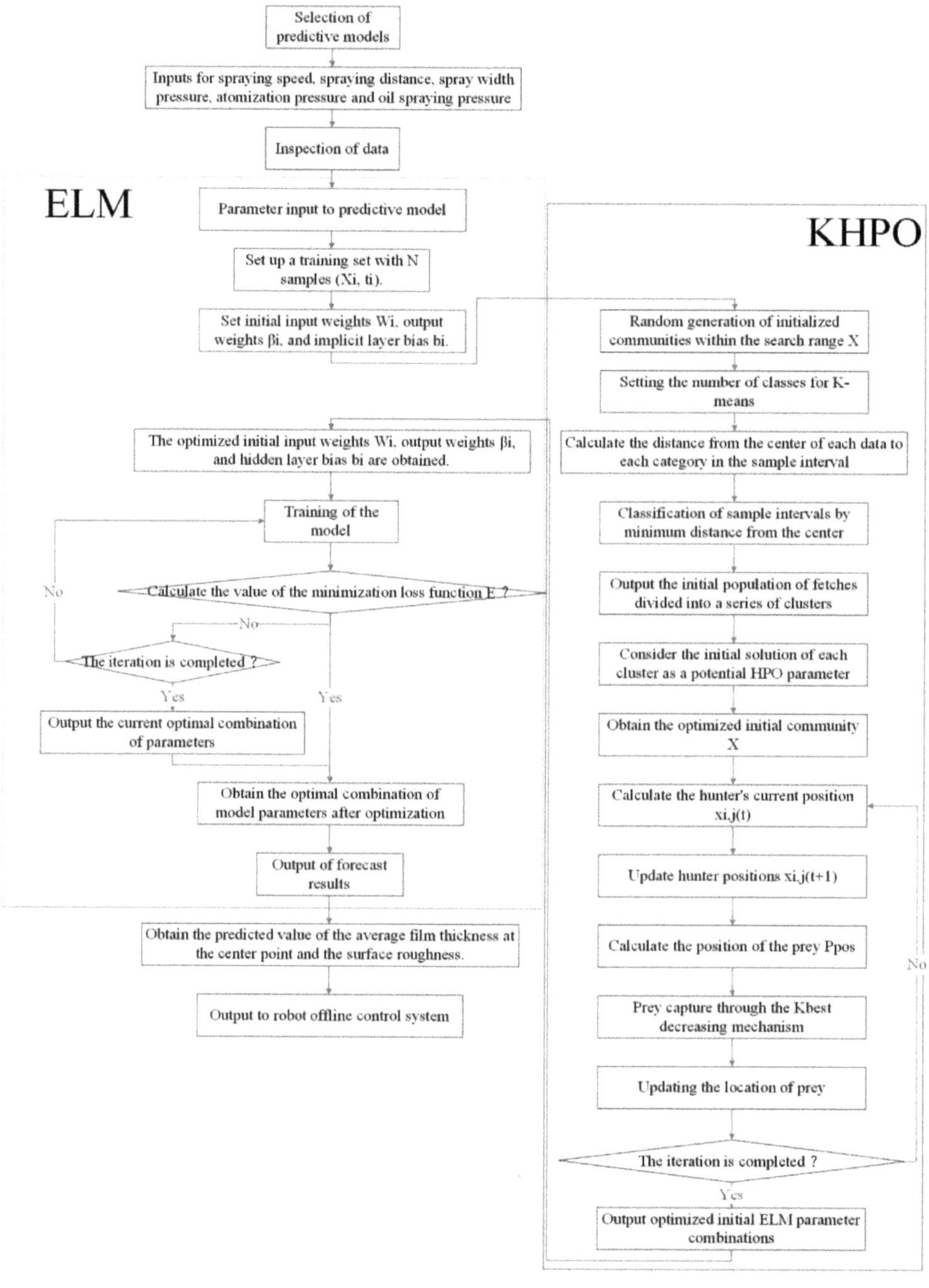

Figure 2. Block diagram of the process of spraying quality prediction algorithm based on multiple spraying parameters.

5. Experimental Methods

5.1. Experimental Settings

The experimental subject comprises ABB IR52 six-jointed robotic arm equipped with the GFA-200 automatic paint spray gun, utilizing the rock GFA-200 as the object of study. The research is conducted within an automatic offline spraying system, incorporating custom fixture devices, water-based paint, a magnetic thickness gauge, and a roughness detector. This configuration forms a robotic automated water-based paint spraying experimental system. The system serves as the foundation for designing pre-experiment data acquisition and validation experiments, ensuring the reliable verification of the spraying quality prediction model, as illustrated in Figure 3.

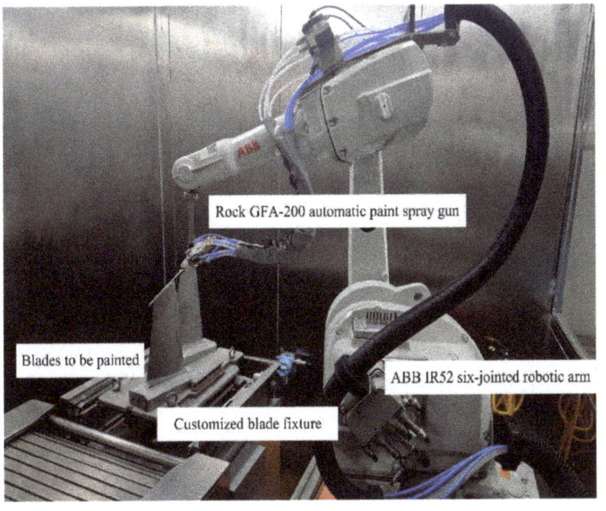

Figure 3. Automated Robotic Waterborne Paint Spraying Experiment System.

5.2. Acquisition of Data

To establish the mapping model capturing the coupling effect between spraying process parameters and spraying quality, this study devises flat waterborne paint spraying experiments involving five input parameters: spraying speed, spraying height, spraying width pressure, atomization pressure, and oil spraying pressure. These experiments, conducted with a magnetic thickness gauge and a roughness detector, evaluate the quality of sprayed plates based on the average film thickness at the center point and surface roughness. The obtained results reveal the impact of various spraying parameters on spraying quality.

The common value ranges for the five input parameters—spraying speed, spraying height, spray width pressure, atomization pressure, and oil spraying pressure—are presented in Table 1 below through the conducted study.

Table 1. Range of values for spraying process parameters.

Parameter	Value
Spraying speed	180 mm/min–280 mm/min
Spraying height	160 mm–230 mm
Spraying width pressure	0.10 MPa–0.17 MPa
Atomizing pressure	0.16 MPa–0.23 MPa
Oil spraying pressure	0.09 MPa–0.13 MPa

In orthogonal experiments, in order to target the study of the effect of the five input parameters, the robot end spraying angle is set at a constant 90° to exclude the interference of the spray gun factors, and at the same time, by ensuring the consistency of the experimental environmental conditions and the material conditions, to exclude the interference of factors such as temperature, humidity, plate, paint, and so on.

To systematically investigate the impact of the five input parameters, a five-factor, four-level orthogonal experiment is designed. The experimental parameter combinations are detailed in Table 2. Employing an isometric sampling approach, the five spraying input parameters are subjected to 3–5 sets of orthogonal experiments within the specified value interval. To mitigate the influence of random errors, three repetitive experiments are conducted to secure more accurate and comprehensive experimental samples and results.

Table 2. Parameter combinations of L16 (4&5) orthogonal experiments.

SN/Parameter	Factor 1: Spraying Speed (mm/min)	Factor 2: Spraying Height (mm)	Factor 3: Spraying Width Pressure (MPa)	Factor 4: Atomizing Pressure (MPa)	Factor 5: Oil Spraying Pressure (MPa)
1	1 (180–190)	1 (160–170)	1 (0.10–0.11)	1 (0.16–0.17)	1 (0.09–0.10)
2	1	2 (180–190)	2 (0.12–0.13)	2 (0.18–0.19)	2 (0.10–0.11)
3	1	3 (200–210)	3 (0.14–0.15)	3 (0.20–0.21)	3 (0.11–0.12)
4	1	4 (220–230)	4 (0.16–0.17)	4 (0.22–0.23)	4 (0.12–0.13)
5	2 (210–220)	1	2	3	4
6	2	2	1	4	3
7	2	3	4	1	2
8	2	4	3	2	1
9	3 (240–250)	1	3	4	2
10	3	2	4	3	1
11	3	3	1	2	4
12	3	4	2	1	3
13	4 (270–280)	1	4	2	3
14	4	2	3	1	4
15	4	3	2	4	1
16	4	4	1	3	2

Upon the completion of the spraying experiments, the collected experimental samples undergo center point average film thickness detection and roughness detection using a magnetic thickness gauge and a roughness detector. The magnetic thickness gauge employed in the spraying experiment is illustrated in Figure 4. Each sample undergoes three repetitive measurements, and the average of these measurements is considered the measured value for the average film thickness at the center point. The roughness tester utilized in the spraying experiment is depicted in Figure 5.

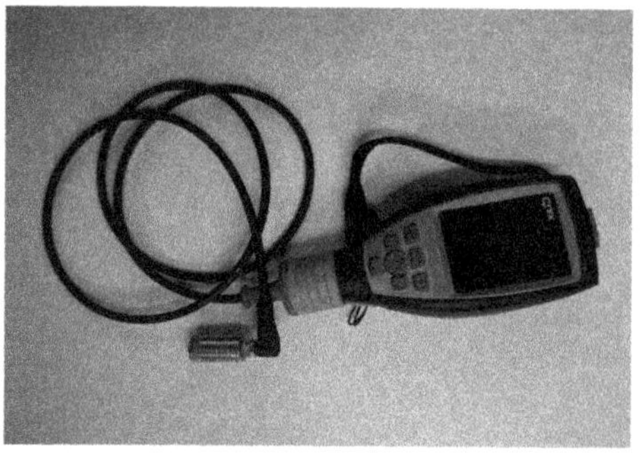

Figure 4. Magnetic Thickness Gauge.

5.3. Validation of Algorithms

In order to verify the reliability of the prediction model, 540 sets of experimental sample data are randomly disrupted in this paper to generate 540 sets of unordered experimental datasets. Then the unordered datasets are divided into 530 sets of training data and 10 sets of testing data. The training datasets are used to construct the KHPO-ELM prediction model, and the test datasets are used to test the reliability of the model.

Figure 5. Roughness Detector.

In this paper, MATLAB software (v. R2022b) is used to build the KHPO-ELM prediction algorithm model for spraying quality of robotic spraying process with multiple inputs and multiple outputs. The initial parameters of the model are set using the experimental method; that is, model testing experiments are conducted for all possible parameters within a reasonable range of values. After several experiments, the model parameters summarized in this paper are shown in Table 3. The evolution curve of the model is depicted in Figure 6.

Table 3. Values of initial parameters of the KHPO-ELM prediction algorithm.

Parameter	Value
Number of neurons in the input layer of the ELM	5
Number of output layer neurons of ELM	2
Number of initial hidden layer neurons of ELM	50
Number of iterations of HPO	500
Population size of HPO	50
Upper and lower bound values of HPO	±1
Contraction factor of HPO	0.1
Initial value of K-means	30

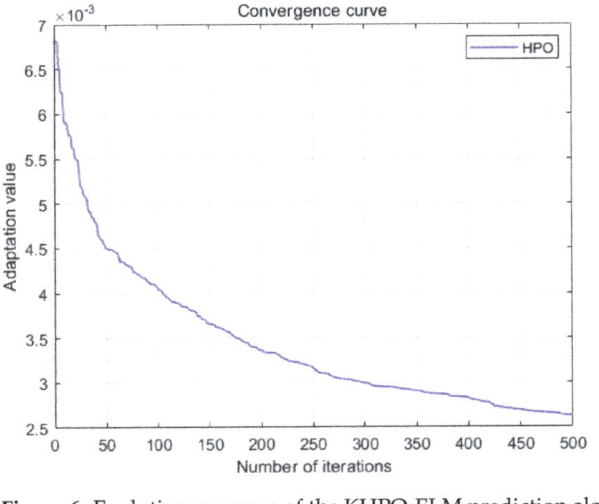

Figure 6. Evolutionary curve of the KHPO-ELM prediction algorithm.

6. Results

To demonstrate and validate the superior performance of the prediction network, two additional prediction models, BP and ELM, are established using the same set of data samples. All three models are input into the identical batch of the robotic automated waterborne paint spraying quality prediction system to assess the feasibility of these prediction models on the same group of test samples. The comparisons of MAE, MSE, RMSE, and MAPE for the prediction data obtained by the KHPO-ELM, ELM, and BP prediction networks are presented in Table 4. Further visual comparisons of predictions are depicted in Figure 7, prediction errors in Figure 8, and the mean absolute error of predictions in Figure 9.

Table 4. Comparison of errors of KHPO-ELM prediction model, ELM prediction model and BP prediction model.

Type	KHPO-ELM		ELM		BP	
	H_{avg}	R_a	H_{avg}	R_a	H_{avg}	R_a
MAE	0.8731	0.0719	2.2946	0.0736	1.7750	0.1563
MSE	0.9483	0.0068	7.6553	0.0077	5.2170	0.0289
RMSE	0.9738	0.0825	2.7668	0.0878	2.2841	0.1701
MAPE	1.4545	9.0026	4.0016	9.1725	2.9918	19.0310

Through validation experiments, the neural network prediction algorithm proves its capability to map coating process parameters to coating quality before the automated robotic coating operation. When compared with ELM and BP neural networks, the KHPO-ELM prediction network exhibits superior prediction accuracy for both the mean film thickness at the center point and surface roughness, eliminating the interference of error factors.

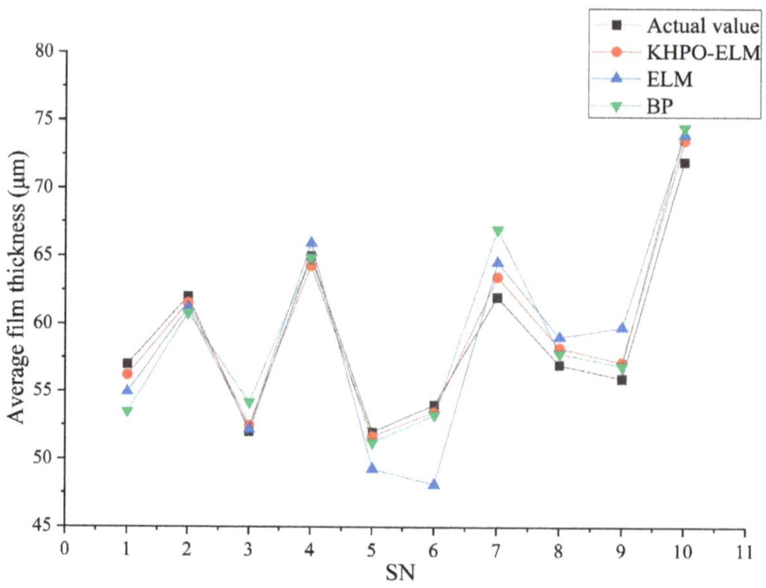

(a) H_{avg}

Figure 7. *Cont.*

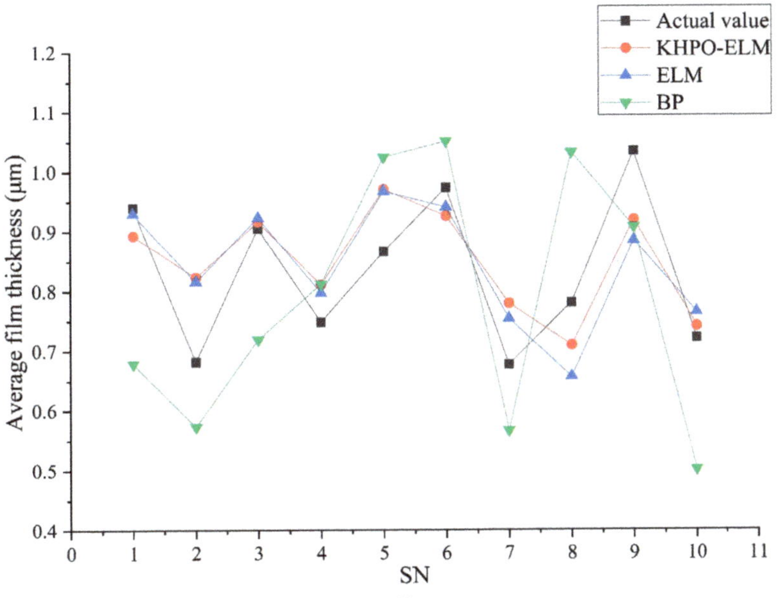

(**b**) R_a

Figure 7. Comparison of predicted values of KHPO-ELM prediction model, ELM prediction model, and BP prediction model.

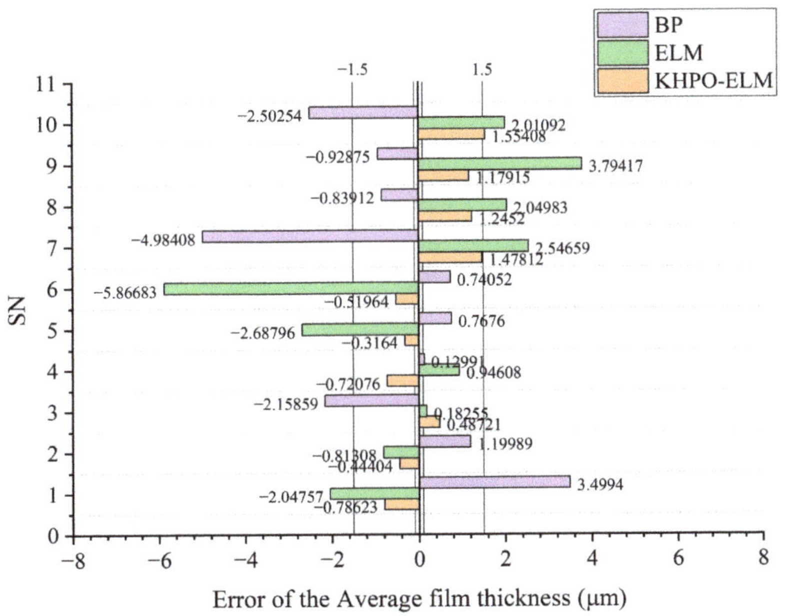

(**a**) H_{avg}

Figure 8. *Cont.*

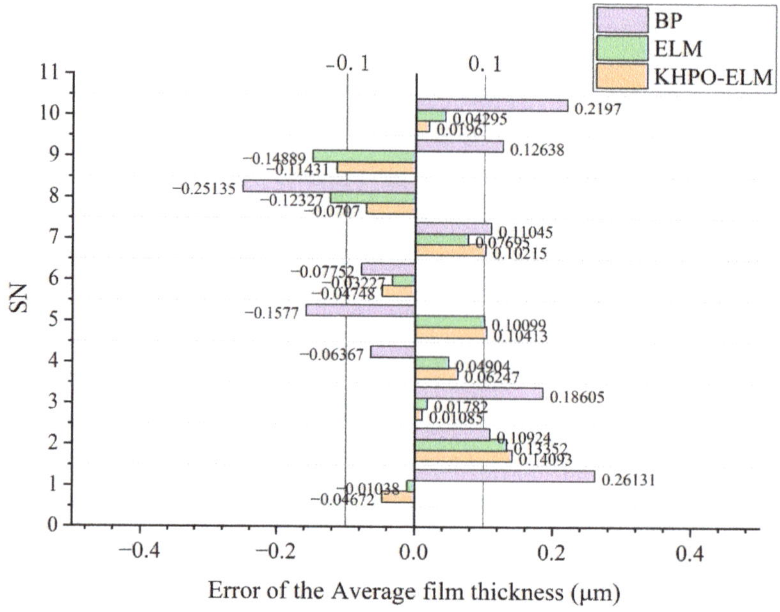

(b) R_a

Figure 8. Comparison of errors of KHPO-ELM prediction model, ELM prediction model, and BP prediction model.

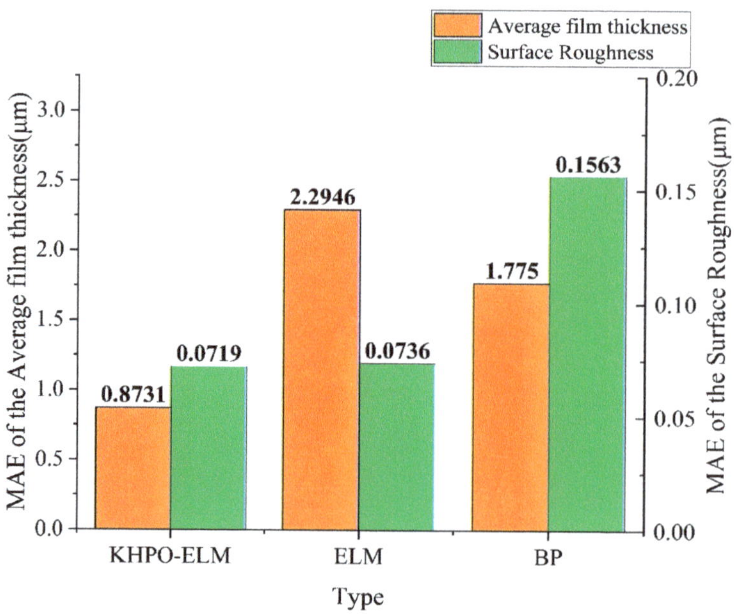

Figure 9. Evolutionary curve of the KHPO-ELM prediction algorithm.

The prediction model excels in forecasting the average film thickness at the center point of the paint film, with its MAE reduced to 0.8731 μm. The errors in the sample test results predominantly fall within ±1.5 μm, meeting the requirements of spraying operations. This reduction amounts to 61.95% and 50.81% when compared to the ELM and BP prediction models, respectively. Similarly, the KHPO-ELM prediction model performs admirably in predicting surface roughness, achieving an absolute average error reduced to 0.0707 μm. The sample test results mostly exhibit errors within ±0.1 μm, meeting the operational requirements. Compared to the ELM and BP prediction models, the error reductions are 2.31% and 54.00%, respectively.

7. Conclusions

(1) The robotic automated waterborne paint spraying process involves predicting the average film thickness at the center of the film and surface roughness, relying on five key spraying process parameters: spraying speed, spraying height, spray pressure, atomization pressure, and spray pressure. Despite the inherent variability in the spraying process, the accumulation of a substantial number of data samples allows for a detailed analysis and summarization of the impact of spraying process parameters during operations. Building upon this, a highly accurate, generalizable, and universal prediction algorithm is proposed, facilitating the rational prediction of spraying process parameters for quality spraying. This algorithm serves as a potent predictor for the application of spraying robots and the implementation of automated coating production lines, demonstrating wide applicability.

(2) In the prediction of coating quality in automated waterborne paint spraying processes by robots, the KHPO-ELM prediction algorithm model demonstrates superior prediction ability and lower prediction errors compared to the ELM and BP models. The MAE, MSE, RMSE, and MAPE indexes of the KHPO-ELM model outperform those of the other two models. The average error of the KHPO-ELM model is 61.95% and 50.81% lower than that of ELM and BP, respectively, ensuring all errors are controlled within approximately ±1.5 μm. Additionally, the average error of the absolute value of surface roughness prediction is 2.31% and 54.00% lower than that of ELM and BP, respectively, with all errors controlled within around ±0.1 μm. This aligns with the quality evaluation requirements of the spraying process.

(3) The core direction for the transformation and upgrading of the machinery industry, particularly in manufacturing processes heavily reliant on manual experience such as welding and grinding, is the substitution of people with robots. The nonlinear mapping model proposed in this paper, based on experimental data and neural networks, stands as a dependable solution to address challenges in this context. This intelligent prediction scheme holds promising applications in the industrial and intelligent transformation of the machinery industry. Primarily, the implementation of such a prediction scheme enhances the intelligence, flexibility, and efficiency of the production process. Additionally, in the realm of human-machine collaboration, this scheme supports machine learning to adapt to human operations and habits, thereby improving the collaborative efficiency between machines and humans. Furthermore, this solution is well suited for future implementation in decision support systems for intelligent mechanical operations. This system not only handles unexpected events adeptly but also optimizes the operational efficiency of mechanical systems. As for the intelligent prediction method, future developmental directions include enhancing the interpretability of the fitted model, improving data security and privacy protection, and advancing multimodal fusion. These aspects are anticipated to contribute to the ongoing evolution and refinement of this intelligent prediction approach.

Author Contributions: Conceptualization, L.L. and X.H.; methodology, X.Z.; software, X.Z.; validation, X.Z., Y.F. and D.Y.; formal analysis, X.Z.; investigation, E.L.; resources, Y.C.; data curation, Y.F.; writing—original draft preparation, L.L. and X.Z.; writing—review and editing, E.L. and X.H.; funding acquisition, Y.C. All authors have read and agreed to the published version of the manuscript.

Funding: This research was funded by the Sichuan University-Yibin School-City Strategic Cooperation Project, grant number 2020CDYB-3.

Data Availability Statement: Data are contained within the article.

Conflicts of Interest: Le Ling, Enpei Liang and Yi Chen are employees of the company.

Appendix A

Table A1. Description of notation.

Notation	Explanation	Unit
H_{avg}	The average film thickness at the center point	μm
R_a	The surface roughness	μm
D	The distribution model obeyed by the quality parameters with spraying	/
V	The spraying speed	mm/min
S	The spraying distance	mm
P_{spray}	The spraying width pressure	MPa
$P_{atomization}$	The atomization pressure	MPa
P_{oil}	The oil spraying pressure	MPa

References

1. Assadi, H.; Kreye, H.; Gärtner, F.; Klassen, T.J.A.M. Cold spraying–A materials perspective. *Acta Mater.* **2016**, *116*, 382–407. [CrossRef]
2. Sarikaya, O. Effect of some parameters on microstructure and hardness of alumina sprayings prepared by the air plasma spraying process. *Surf. Spray. Technol.* **2005**, *190*, 388–393. [CrossRef]
3. Almansoori, N.; Aldulaijan, S.; Althani, S.; Hassan, N.M.; Ndiaye, M.; Awad, M. Manual spray painting process optimization using Taguchi robust design. *Int. J. Qual. Reliab. Manag.* **2021**, *38*, 46–67. [CrossRef]
4. Park, R.M.; Bena, J.F.; Stayner, L.T.; Smith, R.J.; Gibb, H.J.; Lees, P.S. Hexavalent chromium and lung cancer in the chromate industry: A quantitative risk assessment. *Risk Anal. Int. J.* **2004**, *24*, 1099–1108. [CrossRef]
5. Arrais, R.; Costa, C.M.; Ribeiro, P.; Rocha, L.F.; Silva, M.; Veiga, G. On the development of a collaborative robotic system for industrial spraying cells. *Int. J. Adv. Manuf. Technol.* **2021**, *115*, 853–871. [CrossRef]
6. Evjemo, L.D.; Gjerstad, T.; Grøtli, E.I.; Sziebig, G. Trends in smart manufacturing: Role of humans and industrial robots in smart factories. *Curr. Robot. Rep.* **2020**, *1*, 35–41. [CrossRef]
7. Xie, F.; Liu, X.-J.; Wu, C.; Zhang, P. A novel spray painting robotic device for the spraying process in automotive industry. *Proc. Inst. Mech. Eng. Part C J. Mech. Eng. Sci.* **2015**, *229*, 2081–2093. [CrossRef]
8. Pendar, M.R.; Rodrigues, F.; Páscoa, J.C.; Lima, R. Review of spraying and curing processes: Evaluation in automotive industry. *Phys. Fluids* **2022**, *34*, 101301. [CrossRef]
9. Wu, H.; Xie, X.; Liu, M.; Chen, C.; Liao, H.; Zhang, Y.; Deng, S. A new approach to simulate spraying thickness in cold spray. *Surf. Spray. Technol.* **2020**, *382*, 125151.
10. Zhang, Y.; Li, W.; Zhang, C.; Liao, H.; Zhang, Y.; Deng, S. A spherical surface spraying thickness model for a robotized thermal spray system. *Robot. Comput.-Integr. Manuf.* **2019**, *59*, 297–304. [CrossRef]
11. Gleeson, D.; Jakobsson, S.; Salman, R.; Ekstedt, F.; Sandgren, N.; Edelvik, F.; Carlson, J.S.; Lennartson, B. Generating optimized trajectories for robotic spray painting. *IEEE Trans. Autom. Sci. Eng.* **2022**, *19*, 1380–1391. [CrossRef]
12. Gadow, R.; Candel, A.; Floristán, M. Optimized robot trajectory generation for thermal spraying operations and high quality sprayings on free-form surfaces. *Surf. Coat. Technol.* **2010**, *205*, 1074–1079. [CrossRef]
13. Potkonjak, V.; Đorđević, G.S.; Kostić, D.; Rašić, M. Dynamics of anthropomorphic painting robot: Quality analysis and cost reduction. *Robot. Auton. Syst.* **2000**, *32*, 17–38. [CrossRef]
14. Arikan, M.A.S.; Balkan, T. Process simulation and paint thickness measurement for robotic spray painting. *CIRP Ann.* **2001**, *50*, 291–294. [CrossRef]
15. Paturi, U.M.R.; Cheruku, S.; Geereddy, S.R. Process modeling and parameter optimization of surface sprayings using artificial neural networks (ANNs): State-of-the-art review. *Mater. Today Proc.* **2021**, *38*, 2764–2774. [CrossRef]
16. Cao, Y.; Zhao, J.; Qu, X.; Wang, X.; Liu, B. Prediction of Abrasive Belt Wear Based on BP Neural Network. *Machines* **2021**, *9*, 314. [CrossRef]
17. Yang, S.; Nie, W.; Lv, S.; Liu, Z.; Peng, H.; Ma, X.; Cai, P.; Xu, C. Effects of spraying pressure and installation angle of nozzles on atomization characteristics of external spraying system at a fully-mechanized mining face. *Powder Technol.* **2019**, *343*, 754–764. [CrossRef]
18. Wigren, J.; Täng, K. Quality considerations for the evaluation of thermal spray sprayings. *J. Therm. Spray Technol.* **2007**, *16*, 533–540. [CrossRef]

19. Pérez-Ramos, J.D.; Findlay, W.P.; Peck, G.; Morris, K.R. Quantitative analysis of film spraying in a pan coater based on in-line sensor measurements. *Aaps Pharmscitech* **2005**, *6*, E127–E136. [CrossRef]
20. Huang, G.B.; Zhu, Q.Y.; Siew, C.K. Extreme learning machine: Theory and applications. *Neurocomputing* **2006**, *70*, 489–501. [CrossRef]
21. Naruei, I.; Keynia, F.; Molahosseini, A.S. Hunter–prey optimization: Algorithm and applications. *Soft Comput.* **2022**, *26*, 1279–1314. [CrossRef]
22. Krishna, K.; Murty, M.N. Genetic K-means algorithm. *IEEE Trans. Syst. Man Cybern. Part B* **1999**, *29*, 433–439. [CrossRef]

machines

MDPI

Article

Methodology for Rationalization of Pre-Production Processes Using Virtual Reality Based Manufacturing Instructions

Konstantin Novikov, Petr Hořejší *, Jan Kubr, Matěj Dvořák, Miroslav Bednář , David Krákora, Matěj Krňoul * and Michal Šimon

Department of Industrial Engineering and Management, Faculty of Mechanical Engineering, University of West Bohemia in Pilsen, Univerzitni 8, 301 00 Pilsen, Czech Republic; kubrj@fst.zcu.cz (J.K.); dvor1s@fst.zcu.cz (M.D.); bednarm@fst.zcu.cz (M.B.); krakorad@fst.zcu.cz (D.K.); simon@fst.zcu.cz (M.Š.)

* Correspondence: tucnak@fst.zcu.cz (P.H.); mkrnoul@fst.zcu.cz (M.K.); Tel.: +420-608-152-357 (P.H.)

Abstract: This article deals with the rationalization of manufacturing processes within the product life cycle with emphasis on the pre-production phase of production. A new methodology for evaluating the applicability of modern visualization tools in manufacturing processes is presented. This methodology includes a modified Z-score for categorizing manufacturing processes and has been validated by the successful implementation of 10 real projects. Ultimately, the methodology provides a practical decision-making aid for manufacturing companies in deploying such Computer Aided Instruction tools. For the pre-production phase of products and their development, the possibilities of using modern visualization tools to support CAD instructions and assembly instructions are being explored. These modern visualization tools are video tutorials, augmented reality tutorials and virtual reality tutorials. This paper explores the use of these tools for rationalization of pre-production processes. A methodology was designed to select the most appropriate tool for rationalizing process execution in preparation for production. The functionality of the methodology was verified by applying the methodology in industrial practice and subsequent implementation of the recommended solutions. The methodology was validated by testing key combinations that can arise based on the methodology directly in the operations of manufacturing companies. A total of 10 implementations in real production processes were tailored to this study and carried out over 2 years and the functionality of the methodology was confirmed (that consisted also of a new software development). It was found that there is a dependency between the visualization tools chosen to create the instructions in the context of organizational production preparation and the nature of the production processes.

Keywords: assembly instructions; virtual reality; CAD instructions; CAI; VR instructions; production process

1. Introduction

Due to the increasing demands for the development and innovation of new products in the markets, all industrial enterprises must deal with the issue of technical preparation of production. The pre-production phase is an important phase of the product life cycle, as it influences the costs of the other life stages of the product [1,2]. Up to 80% of production costs can be most influenced in the pre-production stages. In these stages, according to [3,4], up to 75% of the various errors that affect product quality, production time, and cost of production also occur. Furthermore, these stages usually involve project-driven activities, which are unique and therefore entail a high level of risk [5]. The technical preparation of production can be influenced by the nature of the workplace and the production process as a whole [6]. The importance of this stage is further enhanced by the high turnover rate and the proportion of foreign workers in industrial enterprises. It is important to address the appropriate training of employees and the form of instructions for new products, as an organized introduction to the process reduces the negative effects of environmental change [7].

The current trend is to move away from paper manuals [8]. Digitized manuals lead to better overall results and a reduction in errors [9]. Some workers may need additional support that a conventional digital manual may not provide [10]. For this reason, interactive digital tutorials are coming to the fore, which may be in the form of video, augmented reality or virtual reality [11]. These tutorials can respond appropriately to the needs of the trainee and improve their overall effectiveness [12].

Video tutorials were found to be beneficial and serve as a useful support tool for practical learning [13]. Several factors determine the quality of a video tutorial such as the instructor's behavior and visual presence, content, scene and other factors [14]. A video recording of the manual assembly process is a simple approach for creating assembly instructions to teach assembly procedures to untrained workers [15]. Workers working from video instructions are more successful in completing the task than workers working from paper instructions [16] and at the same time, workers' confidence in their abilities may increase [17].

Augmented reality (AR) is interactive and combines real and virtual objects [18]. The interactivity brought by this technology is one of the key factors to improve process flexibility and is also a tool to influence the quality and quantity of production [19]. The behavioral effects of AR instructions stood out more in a complex task than in a simple task [20]. Augmented reality tutorials are also suitable for assembling small parts [21] and the difficulty of the tutorial can be adapted to the experience level of the trainee [22]. An alternative use of augmented reality is to display it through a mobile app and project the AR onto the operator's mobile phone display [23]. This restricts the worker's upper limb movement, so the display is often head-mounted (head-mounted system) [24]. With the combination of a head-mounted display and augmented reality, instructors can dynamically place the display and annotations on the physical environment in the worker's field of view during maintenance and repair, thereby instructing workers from anywhere [25]. However, research comparing a video call with an expert to an augmented reality tutorial showed that most interviewees preferred the AR tutorial to a video call with a trainer [26]. The development of augmented reality tutorials is currently quite a lengthy and challenging process [27,28]. Within the service and maintenance, it is possible to observe many global applications that are implemented using an indirect view, i.e., for example, using a tablet or a mobile phone. For direct view applications, for example using semi-transparent glasses (e.g., MS Hololens 2), we face many technological limitations. It seems that the use of glasses that mediate the final image by means of cameras (e.g., Meta Quest 3) may be the right way forward, but here we can face legislative restrictions (from an occupational safety point of view, it is usually not possible to work with opaque glasses). Within the research on the use of augmented reality in service and maintenance, a number of scientific studies confirm that this direction promises great innovations. (e.g., [29–31]).

Virtual reality is computer-generated simulation of an immersive and interactive environment that can be experienced by individuals using specialized hardware, such as VR headsets or goggles. In a virtual reality environment, users are surrounded by three-dimensional (3D) visuals. Virtual reality environments can help increase user creativity levels [32]. The use of virtual reality (VR) to create instructions and assembly procedures has already been addressed in several papers [33,34]. Results have shown that there is a lack of evaluation methods applicable to industrial practice and it is difficult to obtain feedback, although there are methods aimed at learning effects [35]. Furthermore, it was found that virtual reality does not help to acquire motor skills [36] and is more suitable for workflow training [37]. The simplified representation of work steps as well as the lack of multimodal stimuli during VR training negatively affect learning [38]. Thus, it is important to maintain a high level of immersion [39] and training is de-tuned from distracting errors by the trainee [40]. In an experiment comparing paper tutorials, video tutorials, augmented reality, and virtual reality under constrained conditions, it was found that augmented reality led to the most satisfactory task performance [41].

Digital work instructions can be partially generated using artificial intelligence from CAD model BOMs (Bill of Material) [42] or a BOM of a recognized object can be generated [43,44]. Several tests of the automatic generation of augmented reality generated product assembly workflows have been performed, showing that it is possible, albeit with technological limitations, to machine-generate an augmented reality instruction manual [45]. However, in the sources studied, no dependency was found between the use of advanced visualization tools in the context of technical production preparation and the nature of the production processes [46,47] (there are only partial studies that are limited to limited aspects, namely under laboratory conditions, e.g., [48]). This is because manufacturing processes have many parameters (e.g., [49]) that could have a major influence on whether it is appropriate to use modern tools for a particular manufacturing process, or which tool is more suitable for a given process [50]. No methodology has been found that provides recommendations on whether the use of modern tools for instructional design is appropriate under given conditions, or which type of instructional design would be most suitable for a given process [51]. From the literature studied, it was found that the complexity of the different stages of technical preparation for production is increasing, while at the same time the emphasis is on the rapid implementation of this pre-production phase. Companies that have a well-developed pre-production phase are at an advantage [52]. To ensure the efficiency of production execution, modern tools that can work with virtual (physically non-existent) objects can be used [53]. This approach would allow parallel execution of the different stages of technical preparation of production, which would have a positive impact on saving time [54].

Although we have several types of modern visualization tools at our disposal, it is not always clear when these tools can be used for specific production processes in order to make the organizational preparation of production more efficient [55]. At the same time, the question of which type of visualization tools should be selected and possibly combined under predefined conditions is not answered [56]. In practice, then, it seems that companies are pushed into continuously rationalizing their production processes and deploying modern methods, techniques, and tools that are expected to help. However, we often see the opposite effect [57]. Enterprises deploy one of the modern visualization tools without knowing the main added value and limitations of this tool.

Therefore, the main objective of this study is to create a methodology for the rationalization of the implementation of production instructions within the technical preparation of production. In order to achieve this goal, several sub-objectives need to be met:

1. Divide the process of designing the methodology into preparation, decision-making, evaluation and implementation phases;
2. Assess the relationship between the nature of manufacturing processes and visualization tools to rationalization of production execution;
3. Build a criterion function for evaluating the suitability of visualization tools with respect to the nature of production;
4. Define parameters that affect how visualization tools are used;
5. Develop a methodology for selecting a tool that optimally streamlines processes in preparation for production execution, distinguishing between usability and suitability based on external factors;
6. Verification.

The following hypotheses emerged from the main objective and the individual sub-objectives which will be tested:

- There is a dependency between the choice of visualization tools for the creation of instructions for the organizational preparation of production and the nature of the production processes;
- According to the characteristics of the production process, it can be assessed whether the use of modern visualization tools is suitable;
- The proposed methodology can be used to select the appropriate type of visualization tool for a specific manufacturing process.

2. Materials and Methods

The actual research methodology consists of 4 main phases as shown in Figure 1. The individual phases are dealt with completely independently but are logically linked to each other. Each phase contains several steps at the same time. This means that all steps must be gone through in sequence to achieve the expected result. This part of the methodology covers sub-objective 1.

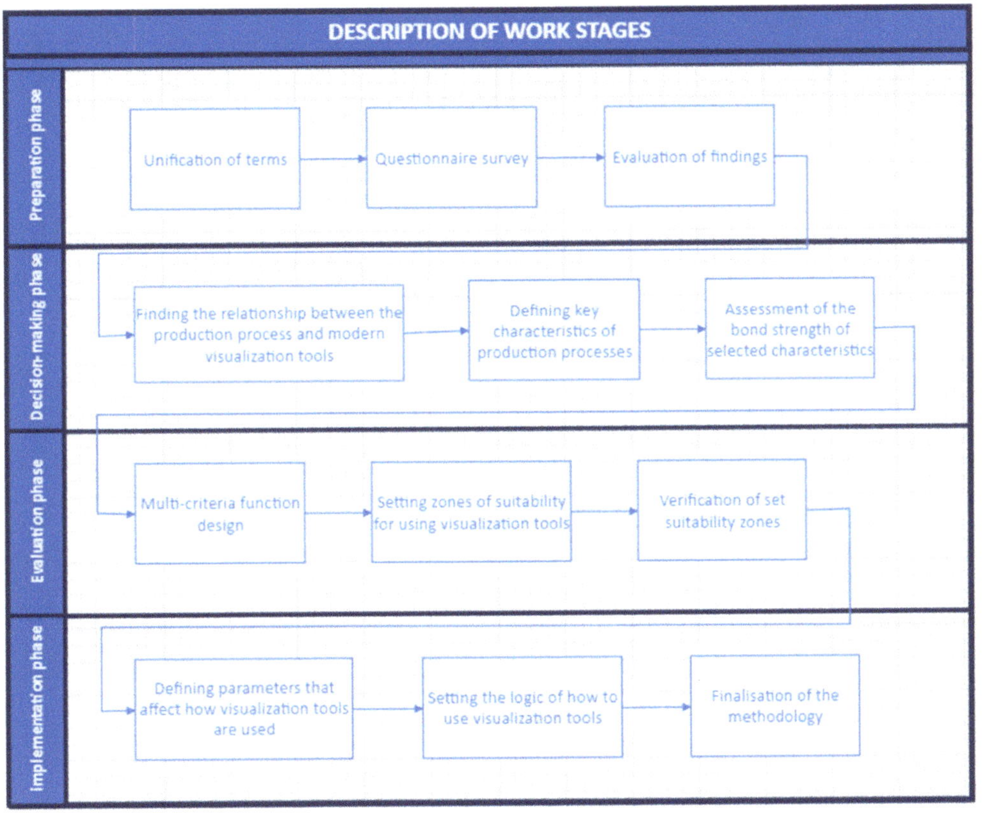

Figure 1. Stages and steps [source: author].

2.1. Preparatory Phase

The preparatory phase is finding out the real state of the given issue in industrial practice. This phase is used to assess the specific problems related to the use of modern visualization tools in the organizational preparation of production. The main objective of this part is to support the theoretical benefits of the research activities with practical benefits.

In cooperation with the Ministry of Labour and Social Affairs of the Czech Republic and the Research Institute of Occupational Safety, a study was designed to assess the relevance of the topic and the contribution of its outputs to industrial practice. The aim of the research was to find the main problem areas related to the use of modern visualisation tools in industrial practice. This includes confirmation of the theoretical and practical benefits of the research and satisfies sub-objective 2 of the methodology. Types of tutorials (modern visualization tools) are divided into:

- Conventional instruments:
  - Paper instructions;

- Modern visualisation tools:
  - Video tutorial;
  - Virtual tutorial;
    - 3D—in virtual reality goggles;
    - 2D—on the display device;
  - Augmented Reality (AR) tutorial;

  How to use the instructions is broken down into:
- Location and interconnection with the production process:
  - Off-site (forward modelling);
  - At the existing workplace;
- Positions within the production process:
  - Stationary (fixed to the workplace);
  - Portable (no fixed link to the location of the production process).

Before approaching the companies, it was necessary to prepare a set of questions to enable a proper understanding of the issues and the potential for further research. The following key questions were selected with the help of experts from the relevant departments:

- Do you use modern visualization tools?
- Are you considering using them?
- Are you considering using them in the organizational preparation of production?
- Do you know when it is appropriate for you to use modern visualization tools?
- How to choose a suitable tool?
- How to use the tool/how to set up the methodology?

2.2. Decision-Making Phase

In this phase, the main objective is to define the PROCESS-TOOL relationship. First, it is necessary to find those characteristics of production processes that influence the decision about the suitability of using modern visualization tools in the organizational preparation of production. This is followed by identifying the key characteristics and assessing the strength of the link. In order to do this, it is necessary to use research methods to design a methodology that can take the above points into account. Define the parameters that influence how visualization tools are used is sub-objective number 4 for the development of the methodology.

Figure 2 represents the work with the individual characteristics of the production processes and the listing of the steps of the decision phase and their logical sequence.

The Likert scale was used to find the relationship between manufacturing processes and modern visualization tools. This measurement technique needs to be redesigned to fit the topic. Most often there are between 5 and 7 options to choose from [58]. The Likert scale makes it possible to determine not only the content of an attitude, but also its approximate strength [59].

In order to use the Likert scale, it is necessary to design an evaluation scale, describing the individual levels with respect to the issue under study. The modified scale consists of 5 levels, where the extreme points are opposite poles. The individual levels are used to evaluate the selected characteristics of the production processes and their impact on the application of modern visualization tools.

A modified scale was created to assess the relationship–see Table 1. For the actual use of the modified scale, important characteristics of the production processes are defined:

- Technical aspect;
- Nature of production;
- Smoothness of production;
- The position of the human in the process;
- Human Resources (Qualifications);

- Impact in case of error;
- Repeatability of production;
- Continuous operation time;
- Production lead time;
- Currency exchange;
- Production and transport dose;
- Step time (operation time/count);
- Methods of transferring components;
- Personnel flows (place attachment);
- Work execution;
- Spatial arrangement;

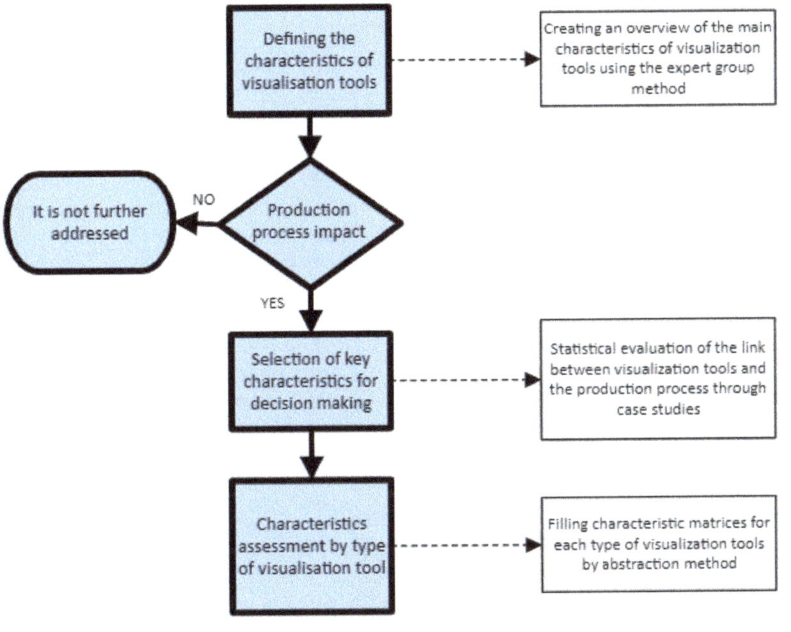

Figure 2. Steps of the decision-making phase [Source: Author].

Table 1. Modified Likert scale [Source: Author].

Rating Scale Modification	
Scale	Levels description
1	The characteristics of the production process have no influence at all on the choice of visualization tools in the organizational preparation of production
2	The characteristics of the production process are only partially related to the choice of visualization tools in the organizational preparation of production
3	The characteristics of the production process have a partial influence on the choice of visualization tools in the organizational preparation of production
4	Characteristics of the production process significantly influence the choice of visualization tools in the organizational preparation of production
5	The characteristics of the production process have a key influence on the choice of visualization tools in the organizational preparation of production

The above characteristics, based on the literature [60–63], will be investigated to determine the impact on the choice of visualization tools and the strength of the correlation. Developing a methodology for selecting a tool that optimally streamlines processes in preparation for production execution, differentiating usability and suitability based on external factors is sub-objective number 5. To this end, an evaluation form was developed.

The next step is to use a modified Likert scale to identify key characteristics of the manufacturing processes that impact the assessment of the suitability and selection of visualization tools. The expert group research method was chosen to accomplish this task. The expert group has 21 members who are qualified representatives from industrial enterprises and research institutions with the following expertise:

- Industrial and process engineers;
- Production managers;
- Developers and programmers;
- Researchers.

2.3. Evaluation Phase

The aim of this phase is to design a multi-criteria function that can assess the suitability of deploying modern visualization tools in the production processes of industrial enterprises. If the suitability of their deployment is proven, it will be further necessary to decide what type of visualization tools will be chosen and how they will be used. The construction of a criterion function for evaluating the suitability of visualization tools with respect to the nature of production is part of sub-objective number 3 for the development of the methodology.

The proposed function will be based on the existing Altman model. The Altman model (also known as the bankruptcy identifier) is indicative of the financial situation of companies. It can predict the financial failure of a firm using statistics. This model was introduced by Prof. Edward Altman and expressed by a function called the Z-score [64].

The Z-score formula is a combination of five indicators on the principle of discriminant analysis which expresses the general notation of the discriminant function [64].

$$Z = 1.2\ X_1 + 1.4\ X_2 + 3.3\ X_3 + 0.6\ X_4 + 1.0\ X_5$$

This is an example of one of the most common formulas, others with different multiples can be used [64].

The ratio indicators contain 5 main values.

X_1 = working capital/firm assets. Simplistically, we can say that a firm in loss can be expected to have liquidity problems as the ratio of current assets to total assets decreases. The problem with this relationship is that although there is a fairly clear relationship between liquidity and solvency, the relationship is significantly skewed in favour of joint stock companies as they have higher net working capital.

X_2 = X2 = retained earnings/assets of the firm. It is not entirely clear whether to count funds from earnings as well, and the literature differs on this point. The weakness of this indicator is that it significantly favours companies that have been in existence longer due to the fact that they have built up sufficient reserves, however, currently in the Czech environment the creation of reserves is not mandatory, which again may affect this indicator. The original idea is that if there are reserves, bankruptcy is less likely, but in an emerging market this weakness is balanced by rapid growth, which the Altman model is unable to account for.

X_3 = X3 = Earnings before interest and taxes (EBIT)/company assets.

X_4 = X4 = Market value of equity (Market capitalization)/foreign capital.

X_5 = X5 = Sales/Assets.

The evaluation allows the classification of the evaluated enterprise into 3 zones (This classification is part of the Altman models described in [64]).

- 3.00 < Z-Safe Zone-Businesses with a high probability of survival

- 1.80 < Z < 2.99-Grey zone-Cannot be clearly identified
- Z < 1.79-Crisis zone-Enterprises in imminent danger of bankruptcy

The Z factor should predict corporate bankruptcies two years in advance, the further into the future we want to predict, the lower the probability of bankruptcy.

In our particular case, we will then look for zones for these ranges:

Conventional tools zone

This zone is given by the resulting modified Z-score value, which is less than 9.5. If we reach these values in the multicriteria function, it is recommended to use conventional tools such as paper-and-pencil. Modern visualization tools would not be effective in this zone.

The grey zone

The grey zone represents a statistically indeterminate area. Thus, if the multicriteria function results in values between 9.5 and 14.0, it means that it is not clearly established whether it would be more appropriate to use conventional or advanced visualization tools. Careful consideration of process, time and financial factors will be required to make a decision in this zone.

Zone of use of modern visualization tools

A multicriteria function result that is greater than or equal to 14.0 represents a zone where it is appropriate and effective to use modern visualization tools. However, this zone does not in itself define which of the modern visualization tools to choose for a particular manufacturing process. It is necessary to assess the parameters of the production process.

2.4. *Implementation Phase*

The aim of this phase is to build on the multicriteria function proposed above. This function, whose result is characterized as a Z-score, is used to assess the suitability of deploying modern visualization tools. If the production process falls within the zone of use of modern visualization tools, it will be necessary to decide further which type of visualization tools will be chosen and how they will be used. Verification is sub-objective number 6. Our work does not deal with the zone of conventional tools or the grey zone. The implementation process is described in detail in the Results section.

3. Results

This chapter consists of a total of four phases:

- Preparatory phase;
- Decision-making phase;
- Evaluation phase;
- Implementation phase.

In each phase, partial results will be presented and evaluated. Finally, a statistical evaluation of the presented combinations will be presented, and a matrix will be created to help determine the most appropriate tool.

3.1. *Preparatory Phase*

The prepared questionnaire was sent to representatives of 100 companies based in the Czech Republic. The intention was to reach a sufficient sample of representatives of industrial enterprises by size:

- Small and micro enterprises defined as having fewer than 50 employees and an annual turnover or balance sheet total of less than EUR 10 million;
- Medium-sized enterprises-these are enterprises which employ fewer than 250 persons and whose annual turnover does not exceed EUR 50 million or whose annual balance sheet total does not exceed EUR 43 million;
- Large enterprises-these are enterprises with more than 250 persons or an annual balance sheet total exceeding EUR 43 million or an annual turnover exceeding EUR 50 million.

In order for the results of the questionnaires to be statistically significant, the minimum threshold for the number of respondents for each type of enterprise was set at 15. The total number of respondents was 64. The small business sector was represented by 17 respondents. Medium enterprises had 19 respondents and large enterprises were represented by 28 respondents.

A total of 64 out of 100 industrial enterprises responded to the questionnaire survey. A complete list of questions and responses can be seen in Table 2. The results are divided into small, medium, and large enterprises.

Table 2. Result of the questionnaire survey [Source: Author].

	Questions	Small-Sized Business		Medium-Sized Business		Big-Sized Business	
		YES	NO	YES	NO	YES	NO
1	Do you use modern visualization tools?	2	15	4	15	8	20
2	Are you considering using them?	9	8	13	6	21	7
3	Are you considering using them in the organizational preparation of production?	4	13	11	8	19	9
4	Do you know when it is appropriate for you to use modern visualization tools?	3	14	3	16	5	23
5	Do you know how to choose the right tool?	2	15	4	15	4	24
6	Do you know how to use the tool / how to set up the methodology?	1	16	2	17	2	26

The most important findings from the table above include:

- Most companies do not have direct experience with the use of modern visualization tools;
- Large and medium-sized enterprises are significantly more interested in deploying modern visualization tools;
- The interest in the use of these tools in the organizational preparation of production prevails among medium and large enterprises;
- Businesses are unable to assess when and under what conditions the use of modern visualisation tools is appropriate for them;
- There is no methodology available to help companies choose the most appropriate tool given the characteristics of their production processes;
- Over 90% of businesses do not know how to use the tools.

The result of the investigation demonstrates the practical applicability of the research and the author's assumptions. At the same time, according to the results, it is possible to specify the area of focus of the proposed methodology:

- Large and medium-sized manufacturing companies;
- Type of production-piece, serial, mass;
- Business sector-industrial enterprises.

The methodology is proposed within the limits listed. However, it can be expected to be appropriate in other circumstances. However, this is not the focus of this paper, but rather it raises the potential for further investigation.

3.2. Decision-Making Phase

In order to set up a suitable algorithm for working with the characteristics of production processes, it is necessary to narrow down the number of relevant parameters. Therefore, those with the lowest weight need to be excluded.

The relevance rules for the values obtained for each characteristic were set as follows:

I. The difference between the arithmetic mean and the modulus is at most |1|;

II. The difference between the arithmetic mean and the median is at most |1|
III. The maximum value of the standard deviation is 1.5;
IV. A relevant association is given by an arithmetic mean value of 2 (inclusive) or higher.

In Figure 3 it can be seen that all the set assumptions regarding the relevance of the obtained data are met (specifically rules I, II and III). This confirms the claim that there is a link between the characteristics of the production processes and the assessment of the suitability of deploying modern visualization tools. The results obtained by the expert group method do not deviate significantly and are within the set tolerance.

According to Rule IV, the characteristics of the production processes are also divided into those that do not influence the assessment of the suitability and choice of visualization tools (arithmetic mean < 2) and those where a link has been identified (arithmetic mean ≥ 2).

In Figure 3 one can further see the interpretation of the obtained data using boxplots, i.e., a graphical visualization of the numerical data using their quartiles. We can notice that the middle part of the diagram is bounded by the 3rd quartile from above and the 1st quartile from below. Between them there is a line defining the median. The boxplots also contain lines extending from the middle part of the diagram perpendicularly up and down, called 'whiskers'. These show the variability of the data below the first and above the third quartile. The outliers are then plotted as separate points. The arithmetic mean is indicated by a cross in the body of each boxplot. A red line symbolizing Rule IV is also drawn through the graph.

In Figure 4 can be seen a selection of the characteristics of the production processes that fulfilled Rule IV. This means that they have the greatest influence on the assessment of the suitability, selection of the type and deployment of visualization tools. Now we need to group these characteristics according to their importance. The essential group will influence both the suitability and the type of visualisation tools. Other groups may only influence the type or method of use of visualization tools.

Selected characteristics of production processes that have a demonstrable PROCESS-TOOL link are further divided into three basic groups. Each of the groups A, B, and C has a different relationship to the assessment of the suitability of visualisation tools.

Selected characteristics:

- Technical aspect;
- Human Resources (Qualifications);
- Impact in case of error;
- Repeatability of production;
- Continuous operation time;
- Step time (operation time/count);
- Personnel flows (place attachment);
- Work execution.

Groups:
A—influences the type and use of visualization tools;
B—influences the type of visualization tools;
C—affects the way visualization tools are used.

The expert group method was used to carry out the sorting, which can be seen in Table 3. Each characteristic was examined separately and then assigned to the group with the best consensus of the expert group. This decision is based on the fulfilment of rules I, II, and III, which confirm the relevance of the data and do not exceed the maximum variance limit.

The selected characteristics are further sorted according to parameter P_x. The distribution shows that two characteristics are in group A (P1 and P4), four are in group B (P2, P3, P5 and P6) and two are in group C (P7 and P8). The strength of binding was not taken into account during the sorting of the characteristics. Its importance will be applied when constructing the multicriteria function.

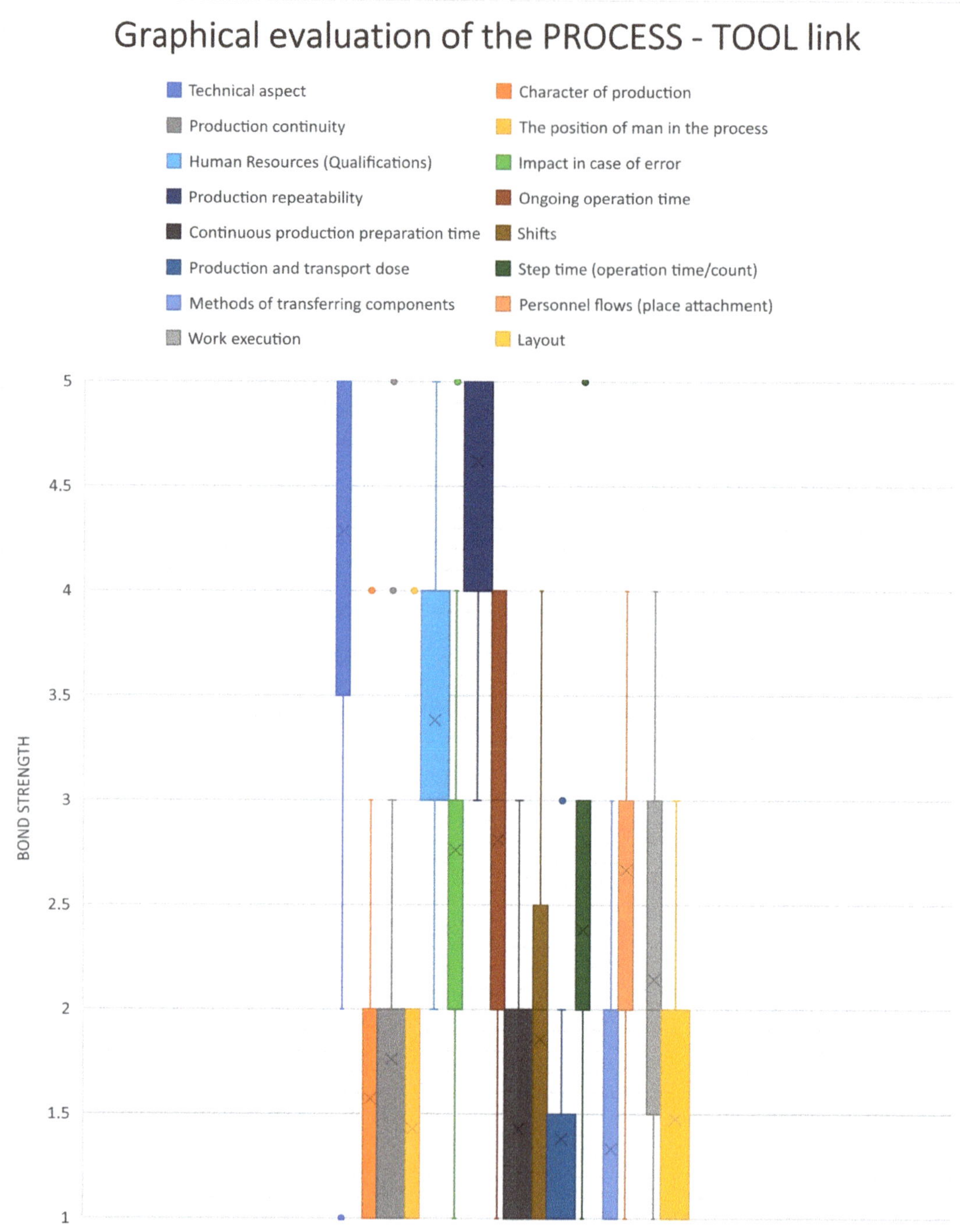

Figure 3. Graphical evaluation of links [Source: Author].

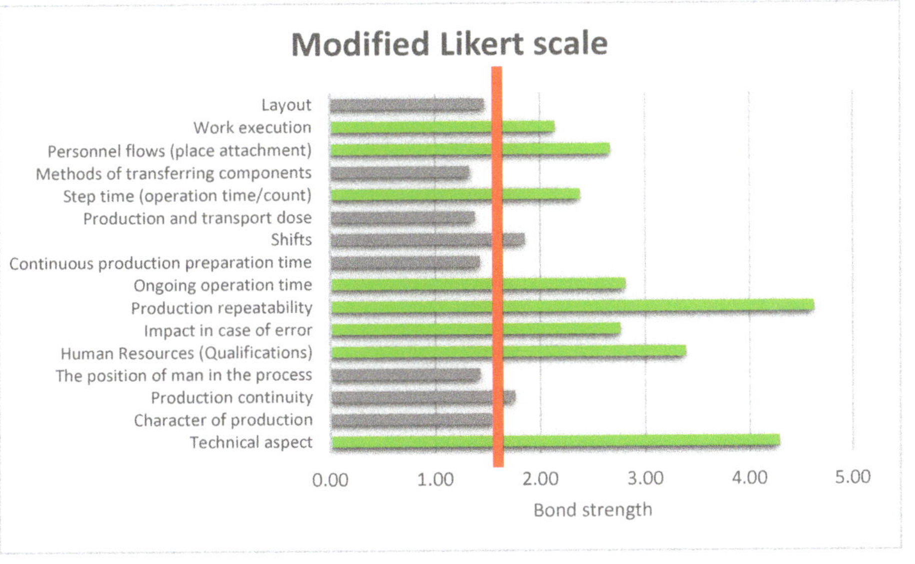

Figure 4. Selection of key characteristics of manufacturing processes [Source: Author].

Table 3. Classification of selected characteristics of production processes into groups [Source: Author].

		Group		
	Bond Strength	**A**	**B**	**C**
P1—Technical aspect	4.29	X		
P2—Human Resources (Qualifications)	3.38		X	
P3—Impact in case of error	2.76		X	
P4—Production repeatability	4.62	X		
P5—Ongoing operation time	2.81		X	
P6—Step time (operation time/count)	2.38		X	
P7—Personnel flows (place attachment)	2.67			X
P8—Work execution	2.14			X

3.3. Evaluation Phase

The proposed multicriteria function is based on the modified Altman model. The characteristics of the production processes, which were classified into groups A and B in the previous chapter, serve as the main indicators. Relevant in this case is the strength of the association of each characteristic and the value of the possible parameter that the characteristic can take on. The formula of the modified Z-factor is designed as follows:

$$Z_{MOD} = S_1 \times Y_1 + S_2 \times Y_2 + S_3 \times Y_3 + S_4 \times Y_4 + S_5 \times Y_5 + S_6 \times Y_6$$

where

From$_{MOD}$—modified multicriteria function score;
S_1—bond strength characteristics Technical aspect;
S_2—strength of the link Human Resources characteristics;
S_3—strength of bond characteristics Impact when fault occurs;
S_4—bond strength characteristics Repeatability of production;
S_5—bond strength characteristics Continuous operation time;
S_6—bond strength characteristics Step time;
Y_1—value of the selected parameter of the characteristic Technical aspect;
Y_2—value of the selected parameter of the Human Resources characteristic;

Y_3—the value of the selected parameter of the characteristic Impact in case of error;
Y_4—value of the selected parameter of the characteristic Repeatability of production;
Y_5—value of the selected parameter of the characteristic Continuous operation time;
Y_6—value of the selected parameter of the Step Time characteristic.

The strength of coupling of process characteristics was quantified in the previous section of the paper. Now we need to define what values each of the characteristics can take. These values must then be numerically ranked according to their impact on the assessment of the suitability of the visualization tools.

The larger the value of the modified Z factor, the more suitable the production process will be for the deployment of advanced visualization tools. Conversely, a low score predicts the deployment of conventional tools as more suitable.

The method of the expert group was also used to determine the values of the parameters of the individual characteristics of the production processes of group A + B. This expert group consisted of fifteen experts from various relevant disciplines. Some of them were contacted in person and others by correspondence. The aim is to clarify the missing specification of the parameters, namely their possibilities and their rating, in order to be able to provide each characteristic its partial score. The partial score of a parameter is calculated as the product of the parameter value and the strength of the association. The sum of the subscores then allows the total modified Z-score to be calculated. This is a prerequisite for categorizing manufacturing processes to interpret the suitability of using modern visualization tools.

In Table 4 the necessary values that can be taken by individual parameters of production processes are added. With these values it is possible to prepare an evaluation and set up zones of suitability for the use of visualisation tools.

The above characteristics are important for obtaining information about production processes. Using the strengths of the relationships between the characteristics of the manufacturing processes and their partial values of each parameter, it is possible to simulate all combinations of the resulting modified Z-score that the manufacturing processes can achieve.

The next step is to set up three main zones that will determine the area of suitability for deploying modern visualization tools.

- Conventional tools zone;
- Grey zone (statistically indeterminate area);
- Zone of use of modern visualization tools.

The 80% agreement for a particular simulated modified Z-score was chosen as the statistically significant value. The simulation results were divided by 0.5 points. The scale is always between 4 and 20 points. The minimum of the simulation was 4.25 points and the maximum was 19.68 points.

For the simulation itself, the possibility of creating different combinations of production process parameters was used. The expert group method was then used to assess whether conventional or modern tools would be appropriate for the score. In Figure 5 it is possible to observe the percentage agreement of the expert group on the suitability of the simulated production processes for the use of visualisation tools. The left side of the graph indicates the agreement in the use of conventional tools. The right side has the agreement in the use of modern visualization tools. In the middle we see a grey area (less than 80% agreement) where it is not possible to clearly determine which of the tools it is appropriate to use.

Table 4. Values of parameters of production process characteristics [Source: Author].

	Parameter Value	Bond Strength S_x	Partial Parameter Y_x Score
P1—Technical aspect			
P11—Greenfield	1	4.29	4.29
P12—Brownfield	0.4		1.71
P2—Human Resources (Qualifications)			
P21—Low	1	3.38	3.38
P22—Medium	0.5		1.69
P23—High	0.1		0.34
P3—Impact in case of error			
P31—Small	0.1	2.76	0.28
P32—Medium	0.6		1.66
P33—High	1		2.76
P4—Production repeatability			
P41—Piece production	0.2	4.62	0.92
P42—Serial production	1		4.62
P43—Mass production	0.5		2.31
P5—Ongoing operation time			
P51—Up to 1 min	0.3	2.81	0.84
P52—1–20 min	0.8		2.25
P53—20–60 min	0.4		1.12
P54—60 min and more	0.1		0.28
P6—Step time (operation time/count)			
P61—Up to 5 s	0	2.38	0.00
P62—6–60 s	1		2.38
P63—60 s and more	0.3		0.71

Based on the reliability test of the simulated Z-scores by the expert group, the individual zones can be specified. An important value is 9.5 points, which defines the boundary between the zone of conventional instruments and the grey zone. The grey zone, on the other hand, is limited by the value of 13.5 points, see Table 5.

Table 5. Zones according to modified Z-score [Source: Author].

$Z_{MOD} < 9.5$	Conventional tools zone
$9.5 \leq Z_{MOD} < 14$	Grey area (statistically indeterminate)
$14 \geq Z_{MOD}$	Modern visualization tools zone

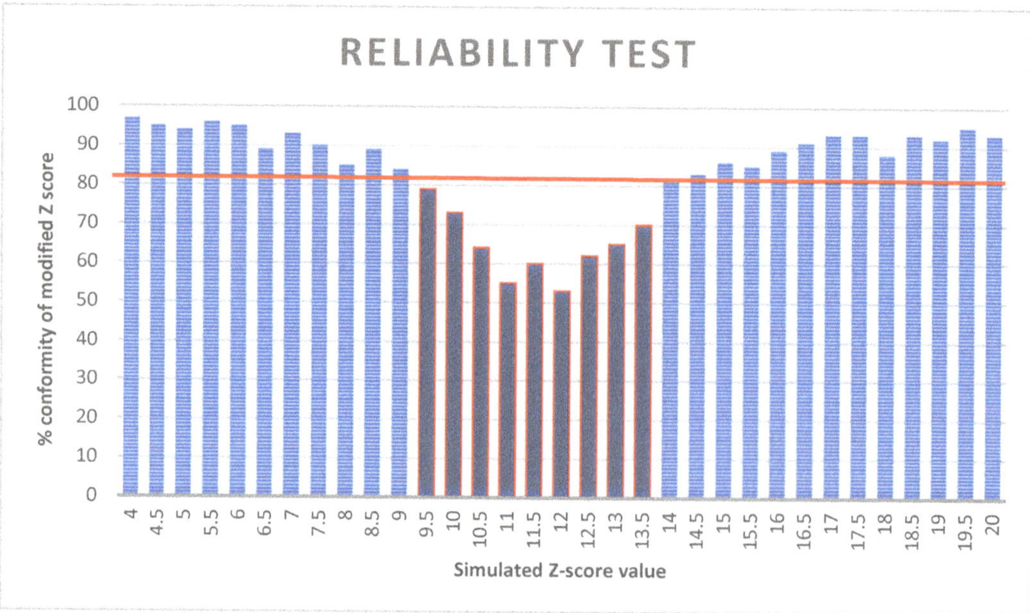

Figure 5. Simulation reliability test [Source: Author].

3.4. Implementation Phase

To be able to decide on the type and use of advanced visualization tools in a zone where the Z-score exceeds 14, we need to define the characteristics that influence the use of these tools. These characteristics were set in Table 3 where we can see their classification into groups A, B, and C. The groups that influence the type and way of using modern visualization tools are A and C. The characteristics belonging to groups A or C include:

- P1—Technical aspect, can take values:
  - P11—Greenfield;
  - P12—Brownfield;
- P4—Repeatability of production, can take values:
  - P41—Unit;
  - P42—Serial;
  - P43—Bulk;
- P7—Personnel flows (place attachment), may take values:
  - P71—Above 80%;
  - P72—Less than 80%;
- P8—Work execution, can take values:
  - P81—Parallel manual work up to 50%;
  - P82—Parallel manual work over 50%.

P7 indicates a person's attachment to the workplace. A worker who does not rotate between workplaces and stays only at one of his/her own will reach the 100% workplace attachment threshold [65]. P8 indicates the amount of manual work performed in a given time and [66]. Each characteristic can take on different values. These values will be further sorted as the letter y for the parameter P_{xy}. Hence, it will be necessary to first compile all possible combinations of these four process characteristics. This results in a total of 24 possible combinations of the selected parameters shown in Table 6. Within these combinations, it is now necessary to evaluate which type and method of using modern visualization tools will be suitable for each option.

Table 6. Possible combinations of selected parameters [Source: Author].

P_{11}	$P_{11}P_{41}$	$P_{11}P_{41}P_{71}$	$P_{11}P_{41}P_{71}P_{81}$ $P_{11}P_{41}P_{71}P_{82}$
		$P_{11}P_{41}P_{72}$	$P_{11}P_{41}P_{72}P_{81}$ $P_{11}P_{41}P_{72}P_{82}$
	$P_{11}P_{42}$	$P_{11}P_{42}P_{71}$	$P_{11}P_{42}P_{71}P_{81}$ $P_{11}P_{42}P_{71}P_{82}$
		$P_{11}P_{42}P_{72}$	$P_{11}P_{42}P_{72}P_{81}$ $P_{11}P_{42}P_{72}P_{82}$
	$P_{11}P_{43}$	$P_{11}P_{43}P_{71}$	$P_{11}P_{43}P_{71}P_{81}$ $P_{11}P_{43}P_{71}P_{82}$
		$P_{11}P_{43}P_{72}$	$P_{11}P_{43}P_{72}P_{81}$ $P_{11}P_{43}P_{72}P_{82}$
P_{12}	$P_{12}P_{41}$	$P_{12}P_{41}P_{71}$	$P_{12}P_{41}P_{71}P_{81}$ $P_{12}P_{41}P_{71}P_{82}$
		$P_{12}P_{41}P_{72}$	$P_{12}P_{41}P_{72}P_{81}$ $P_{12}P_{41}P_{72}P_{82}$
	$P_{12}P_{42}$	$P_{12}P_{42}P_{71}$	$P_{12}P_{42}P_{71}P_{81}$ $P_{12}P_{42}P_{71}P_{82}$
		$P_{12}P_{42}P_{72}$	$P_{12}P_{42}P_{72}P_{81}$ $P_{12}P_{42}P_{72}P_{82}$
	$P_{12}P_{43}$	$P_{12}P_{43}P_{71}$	$P_{12}P_{43}P_{71}P_{81}$ $P_{12}P_{43}P_{71}P_{82}$
		$P_{12}P_{43}P_{72}$	$P_{12}P_{43}P_{72}P_{81}$ $P_{12}P_{43}P_{72}P_{82}$

Type and usage logic settings

Each sub-combination of process parameters will be evaluated to assess the suitability of the type and method of use of the visualisation tools. This is necessary to disentangle the relationships and interdependencies of the individual parameters. As in Figure 3 the expert group method was used. The aim is to find the most suitable option (type and method of using modern visualization tools) for each sub-combination. Before the actual evaluation of the variants, rules were set. The relevance rules of the obtained values for each combination were set as follows:

I. The suitable variant is given by the modus function;
II. Value modus > 10 (out of 21 ratings);
III. The minimum Z-score must be at least 14 (rule from the previous phase).

Parameter P1 was selected as the most suitable for sorting combinations. Therefore, all 12 combinations of parameter P11 were evaluated first. These are all the possibilities related to a site that is in the design phase and does not yet exist in reality. The result is:

- P11P41P71P81—the suitable variant is VR 3D selected 15 times, rule II has been fulfilled;
- P11P41P71P82—suitable variant is VR 3D selected 14 times, rule II was met;
- P11P41P72P81—the suitable variant is VR 3D selected 16x, rule II has been fulfilled;
- P11P41P72P82—suitable variant is VR 3D selected 11 times, rule II was fulfilled;
- P11P42P71P81—the suitable variant is VR 3D selected 15 times, rule II has been fulfilled;
- P11P42P71P82—the suitable variant is VR 3D selected 16 times, rule II was fulfilled;
- P11P42P72P81—the suitable variant is VR 3D selected 12 times, rule II was fulfilled;
- P11P42P72P82—the suitable variant is VR 3D selected 18 times, rule II was fulfilled;
- P11P43P71P81—suitable variant is VR 3D selected 14 times, rule II was met;
- P11P43P71P82—suitable variant is VR 3D selected 14 times, rule II was met;

- P11P43P72P81—suitable variant is VR 3D selected 13 times, rule II was met;
- P11P43P72P82—the suitable variant is VR 3D selected 15 times, rule II was fulfilled.

As can be seen in all combinations met the set rules and it was thus clearly determined which type of visualization tool is suitable for specific combinations. Related to this is the choice of the method of use, which was also part of the evaluation. Subsequently, the evaluation was performed by combining the P12 parameter. These are options related to existing sites.

The following results are shown:

- P12P41P71P81—does not meet the minimum required Z-score (Rule III);
- P12P41P71P82—does not meet the minimum required Z-score (Rule III);
- P12P41P72P81—does not meet the minimum required Z-score (Rule III);
- P12P41P72P82—does not meet the minimum required Z-score (Rule III);
- P12P42P71P81—suitable variant is VR 2D stationary selected 14 times, rule II was met;
- P12P42P71P82—suitable variant is VR 2D stationary selected 15 times, rule II was met;
- P12P42P72P81—the suitable variant is VR 2D portable selected 16x, rule II was met;
- P12P42P72P82—suitable variant is AR portable selected 11 times, rule II was met;
- P12P43P71P81—suitable variant is Video stationary selected 12 times, rule II was met;
- P12P43P71P82—suitable variant is Video stationary selected 14 times, rule II was fulfilled;
- P12P43P72P81—suitable variant is VR 2D portable selected 15 times, rule II was met;
- P12P43P72P82—suitable variant is VR 2D stationary selected 12x rule II was met.

The results of the P12 combinations also meet the required rules and are clearly statistically demonstrable. The collected results are used to build a matrix of combinations for the use zone of modern visualization tools, where the processes will be assigned a type and use.

Matrix of combinations of process and application characteristics

Based on the results obtained above (specifically in Tables 7 and 8), it is now possible to construct a matrix of all possible combinations of manufacturing process characteristics and types of modern visualization tools, including how they are used. We only consider those characteristics that have an impact on the type and use of visualisation tools, i.e., belong to groups A or C.

Thus, Table 9 is a representation of the multidimensional matrix containing a complete list of combinations and suitable variants according to the proposed methodology. However, some combinations could not be specified as they do not reach the minimum modified Z-score according to the multicriteria function. Thus, we now have a completed methodology for rationalizing the execution of manufacturing processes in the context of organizational production preparation. Thanks to the proposed methodology, we are able to assess whether a particular manufacturing process is suitable for modern visualization tools. If so, this methodology can be used to select the type and method of using the visualization tool according to the defined characteristics of the production process.

The proposed methodology, which uses the above mathematical apparatus, therefore allows us to recommend a suitable visualization tool-Paper, Video (Stationary/Portable), VR 3D, VR 2D (portable/stationary), AR (portable/stationary)-based on the input, which are the parameters of the specified characteristics of the production processes, which were limited to six (see Table 4). Due to the fact that the above limitations could be found, it is thus possible to perform validation on a limited number of combinations. Validation will be performed for these combinations. Thus, a testbed will be developed for each type of tutorial.

Table 7. Evaluation of P11 combinations [Source: Author].

	Statistical Evaluation of P11 Combinations										
	Paper	Video		VR 3D	VR 2D		AR		Summary	Modus	Rule Fulfilment (Modus Value > 10)
		Stationary	Portable		Stationary	Portable	Stationary	Portable			
P11P41P71P81	0	0	0	15	2	0	0	4	21	VR 3D	YES
P11P41P71P82	0	0	0	14	1	4	0	2	21	VR 3D	YES
P11P41P72P81	0	0	0	16	4	0	0	1	21	VR 3D	YES
P11P41P72P82	0	0	0	11	5	3	0	2	21	VR 3D	YES
P11P42P71P81	0	0	0	15	2	4	0	0	21	VR 3D	YES
P11P42P71P82	0	0	0	16	1	4	0	0	21	VR 3D	YES
P11P42P72P81	0	0	0	12	3	3	0	3	21	VR 3D	YES
P11P42P72P82	0	0	0	18	0	1	0	2	21	VR 3D	YES
P11P43P71P81	0	0	0	14	5	1	0	1	21	VR 3D	YES
P11P43P71P82	0	0	0	14	3	2	0	2	21	VR 3D	YES
P11P43P72P81	0	0	0	13	3	3	0	2	21	VR 3D	YES
P11P43P72P82	0	0	0	15	5	1	0	0	21	VR 3D	YES

Table 8. Evaluation of P12 combinations [Source: Author].

	Statistical Evaluation of P11 Combinations										
	Paper	Video		VR 3D	VR 2D		AR		Summary	Modus	Rule Fulfilment (Modus Value > 10)
		Stationary	Portable		Stationary	Portable	Stationary	Portable			
P12P41P71P81										Unspecified	NO
P12P41P71P82										Unspecified	NO
P12P41P72P81										Unspecified	NO
P12P41P72P82										Unspecified	NO
P12P42P71P81	0	2	0	0	14	0	5	0	21	VR 2D stationary	YES
P12P42P71P82	0	2	0	0	15	1	3	0	21	VR 2D stationary	YES
P12P42P72P81	0	0	2	0	1	16	0	2	21	VR 2D portable	YES
P12P42P72P82	0	0	2	0	0	8	0	11	21	AR portable	YES
P12P43P71P81	0	12	0	0	7	0	2	0	21	Video stationary	YES
P12P43P71P82	0	14	0	0	6	0	0	1	21	Video stationary	YES
P12P43P72P81	0	0	4	0	1	15	0	1	21	VR 2D portable	YES
P12P43P72P82	0	0	2	0	2	12	0	5	21	VR 2D stationary	YES

Table 9. Matrix of combinations of production process characteristics [Source: Author].

P_{11}	$P_{11}P_{41}P_{71}P_{81}$	VR 3D
	$P_{11}P_{41}P_{71}P_{82}$	VR 3D
	$P_{11}P_{41}P_{72}P_{81}$	VR 3D
	$P_{11}P_{41}P_{72}P_{82}$	VR 3D
	$P_{11}P_{42}P_{71}P_{81}$	VR 3D
	$P_{11}P_{42}P_{71}P_{82}$	VR 3D
	$P_{11}P_{42}P_{72}P_{81}$	VR 3D
	$P_{11}P_{42}P_{72}P_{82}$	VR 3D
	$P_{11}P_{43}P_{71}P_{81}$	VR 3D
	$P_{11}P_{43}P_{71}P_{82}$	VR 3D
	$P_{11}P_{43}P_{72}P_{81}$	VR 3D
	$P_{11}P_{43}P_{72}P_{82}$	VR 3D
P_{12}	$P_{12}P_{41}P_{71}P_{81}$	unspecified
	$P_{12}P_{41}P_{71}P_{82}$	unspecified
	$P_{12}P_{41}P_{72}P_{81}$	unspecified
	$P_{12}P_{41}P_{72}P_{82}$	unspecified
	$P_{12}P_{42}P_{71}P_{81}$	VR 2D stationary
	$P_{12}P_{42}P_{71}P_{82}$	VR 2D stationary
	$P_{12}P_{42}P_{72}P_{81}$	VR 2D portable
	$P_{12}P_{42}P_{72}P_{82}$	AR portable
	$P_{12}P_{43}P_{71}P_{81}$	Video stationary
	$P_{12}P_{43}P_{71}P_{82}$	Video stationary
	$P_{12}P_{43}P_{72}P_{81}$	VR 2D portable
	$P_{12}P_{43}P_{72}P_{82}$	VR 2D stationary

4. Validation of the Methodology

Here we describe the verification of the functionality of the proposed methodology through the deployment of the methodology in industrial practice and the subsequent implementation of the recommended solutions. The aim is to validate the methodology by testing the key combinations that can be created based on the methodology directly in the real operation of manufacturing companies. In order to effectively validate the proposed methodology on a larger number of combinations, it is necessary to develop software support that will facilitate filling of the values into the multicriteria function and allow the selection of the appropriate type and method of using visualization tools from a predefined matrix of combinations.

For the verification itself, a number of new in-house applications have been developed, tailored to this study, but mostly deployed in the actual company. Examples of these environments can be seen in the following subsections.

Creation of software support

Software support must respect all the knowledge gained as described in the previous sections. First of all, it is necessary to link the characteristics of the production processes with the assessment of suitability for modern visualisation tools using a multi-criteria function. The output is then a Z-score, which can be used to assess whether the manufacturing process is suitable or not. If the process is suitable, the selection of the correct type respecting the matrix of all combinations created must follow. Finally, these combinations will also allow the selection of the correct application.

In Figure 6 the initial menu of the created software support can be seen, where the blue fields are for adding and selecting from a predefined list. These are selected important characteristics of the production processes that influence the choice of visualization tools.

All blue items must be pre-populated in order to calculate the subsequent modified Z-score. The green fields are generated automatically based on predefined algorithms.

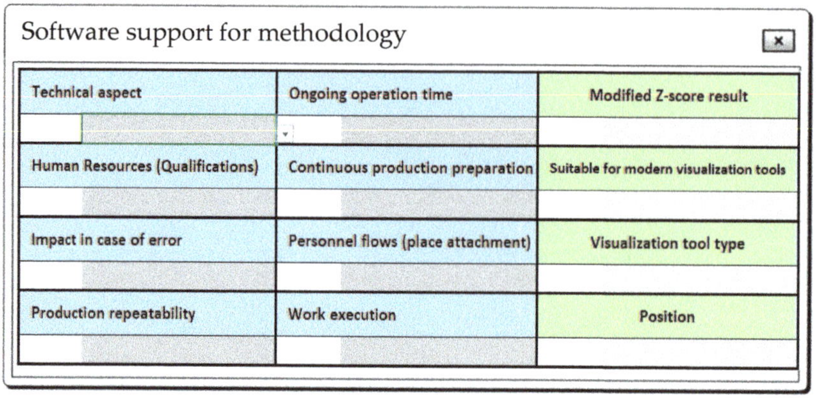

Figure 6. Sample software support [Source: Author].

- Modified Z-score result-calculates the specific value of the modified Z-score according to the proposed methodology;
- Suitable for modern visualization tools-if the Z-score is 14 or higher, this field returns YES (if less than 14, it returns NO);
- Visualization tool type-if YES in the field above, this field returns the specific type of visualization tool according to the combination matrix;
- Position-if the process is suitable for modern visualization tools, it returns the usage method (otherwise it remains empty).

4.1. *Verification of the Combination P11P42P71P82*

In order to verify the combination P11P42P71P82, it was necessary to find a suitable manufacturing process that would not only meet a sufficient value of the resulting modified Z-score, but also match the set parameters of this combination. In Table 10 a specific manufacturing process from an automotive company can be seen.

Table 10. Case Study P11P42P71P82 [Source: Author].

Technical aspect	Ongoing operation time	Modified Z-score result
Greenfield	20–60 min	15.76
Human Resources (Qualifications)	Continuous production preparation	Suitable for modern visualisation tools
Medium	6–60 s	YES
Impact in case of error	Personnel flows (place attachment)	Type of visualisation tool
Medium	Over 80%	VR 3D
Production repeatibility	Work execution	Position
Serial production	Parallel manual work over 50%	-

According to the proposed methodology, the process is suitable for the deployment of modern visualization tools. It is also recommended to use the VR 3D type. For this combination, a standalone VR headset was used, which allows full immersion and interaction with selected elements of the production process. A sample implementation can be seen in the images below. The whole process of the demonstration can be divided into two parts. The first part is purely manual and the second part involves working with automated production equipment as shown in Figure 7.

The demonstration shows a visualisation of some of the steps in the production process. Specifically, this involves removing the prepared parts from the conveyor and then manually assembling the other components, including subsequent inspection and confirmation of the completion of this part with a special button.

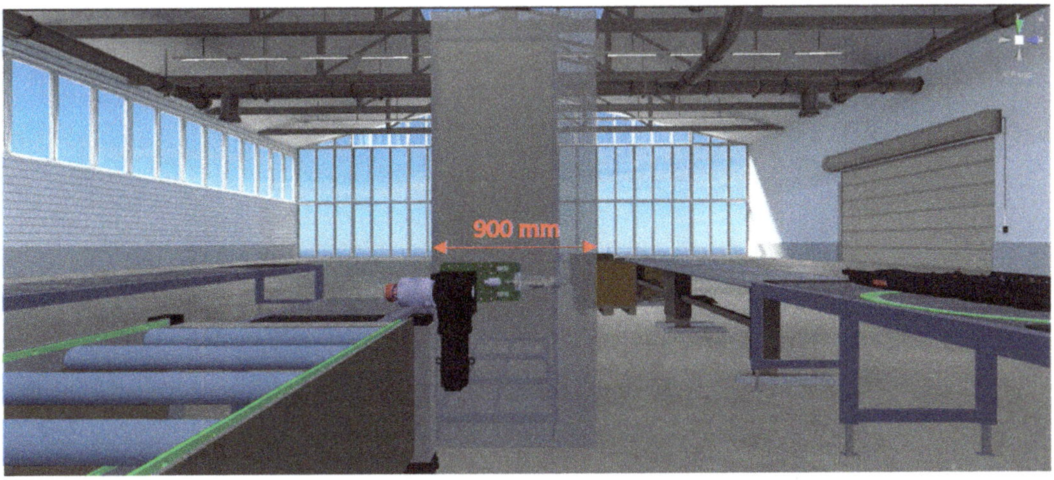

Figure 7. Combination P11P42P71P82 [Source: Author].

Figure 8 shows the second part of the production process. Here, a worker removes the parts from the shelf and mounts them into the production equipment. Next, the rails are set up and a visual inspection takes place. After pressing the device button, the machine part of the work follows.

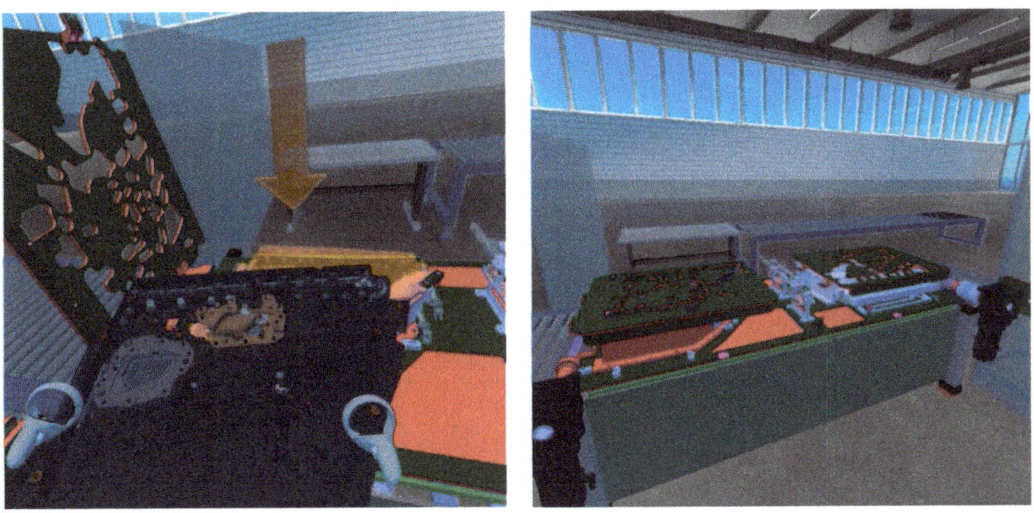

Figure 8. Sample implementation of the P11P42P71P82 combination [Source: Author].

The combination and its suitability was assessed after deployment in industrial practice by an expert group and representatives of the selected company. These were representatives of the technological production preparation department and selected production operators. As a result of the validation, the suitability and type of the developed visualisation tool was demonstrated.

4.2. Verification of the Combination P11P42P72P81

Verification of the P11P42P72P81 combination was carried out on the production process according to the definition of the required parameters. A sufficient modified Z-score was also necessary. In Table 11 can be seen a selected and compliant manufacturing process from a company focused on the production of components for the aerospace industry.

Table 11. Case Study P11P42P72P81 [Source: Author].

Technical aspect	Ongoing operation time	Modified Z-score result
Greenfield	60 min and more	14.35
Human Resources (Qualifications)	Continuous production preparation	Suitable for modern visualisation tools
Medium	60 s and more	YES
Impact in case of error	Personnel flows (place attachment)	Type of visualisation tool
High	Up to 80%	VR 3D
Repeatability of production	Work execution	Position
Serial production	Parallel manual work up to 50%	-

According to the proposed methodology, the suitability for deployment of modern visualization tools was confirmed. It was recommended to use the VR 3D type, as can be seen in the implementation example Figure 9. The visualization was prepared for standalone virtual reality glasses. Again, full interaction with important elements of the process is enabled. Other elements that do not directly interfere with the process are left static without the possibility of interaction.

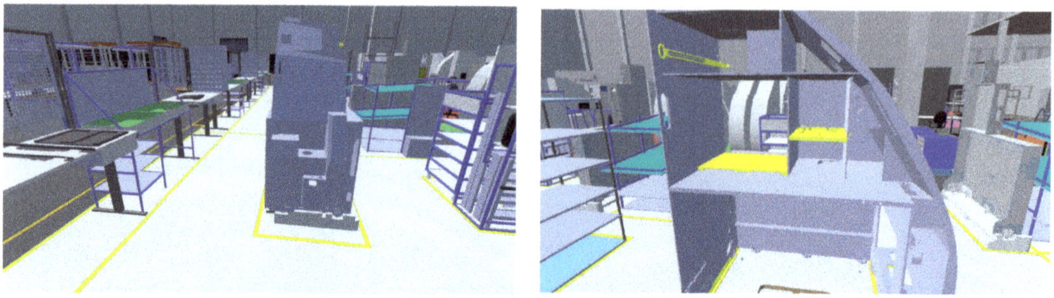

Figure 9. Sample implementation of the P11P42P72P81 combination [Source: Author].

In the first step of the demonstration, the necessary components and tools need to be defined and prepared to successfully complete the task. These components are placed on the table marked in green. Until the preparation is complete, the worker cannot proceed to the next step.

After successful preparation, the individual components are assembled into a larger unit by hand according to a pre-set procedure. Here the worker uses several tools.

This combination was again tested after deployment in industrial practice.

4.3. Verification of the Combination P11P43P71P82

The P11P43P71P82 combination was validated on a production process with a modified Z-score of 14.57. Again, this was an emerging workstation in the design phase, which is specific to high repeatability of production and fast time steps, as can be seen in Table 12.

Using the methodology, it was recommended to use the VR 3D type. The visualization was prepared for a stationary VR headset, as it is commonly used in the company (see Figure 10). The VR headset is equipped with a set of controllers that provide the worker's interactions with the VR environment.

Using the visualization tool, it was possible to simulate the future production process, which consists of five jobs that interact with each other. It is therefore important to synchronise the production steps on all five workstations. VR 3D allowed for the efficient standardization of the individual process steps while aligning all the workstations so that downtime would not occur due to large differences in operation times between interconnected workstations. The speed of production is determined by the slowest link-the workstation.

Table 12. Case Study P11P43P71P82 [Source: Author].

Technical aspect	Ongoing operation time	Modified Z-score result
Greenfield	1–20 min	14.57
Human Resources (Qualifications)	Continuous production preparation	Suitable for modern visualisation tools
Medium	6–60 s	YES
Impact in case of error	Personnel flows (place attachment)	Type of visualisation tool
Medium	Over 80%	VR 3D
Repeatability of production	Work execution	Position
Mass production	Parallel manual work over 50%	-

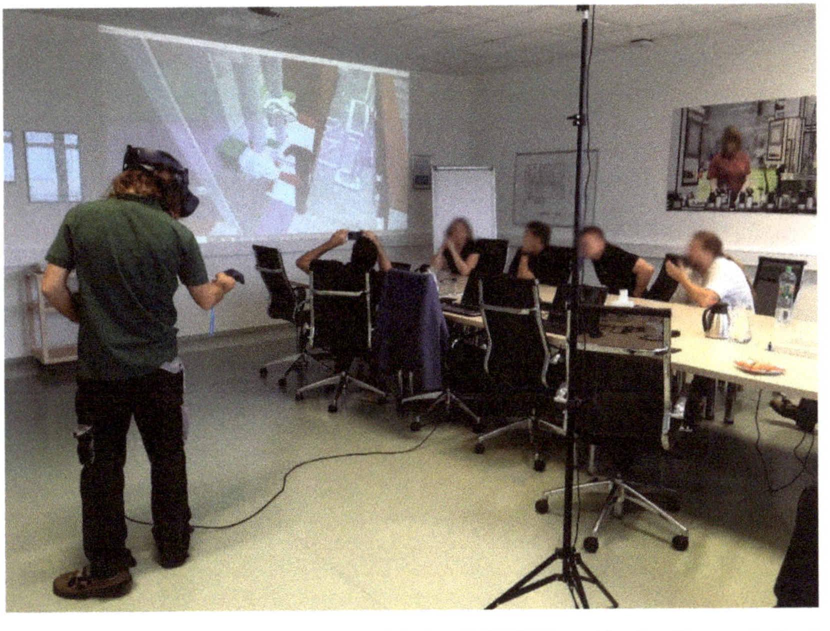

Figure 10. Example of the implementation of the P11P43P71P82 combination [Source: Author].

In Figure 11 you can see the created VR shared workspace environment. Each workspace is Interactable in VR. These workstations are used to load parts into a jig located inside the production facility. The stacking is performed in a pre-set sequence and a clearly defined final position for stacking.

The verification was carried out together with the technical training department, production workers and the expert group. As a result, it was confirmed that the solution was correctly selected according to the methodology.

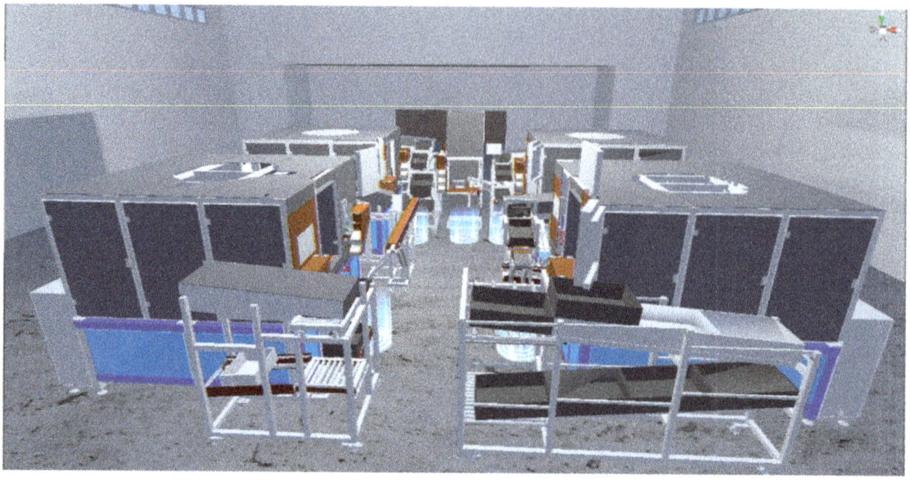

Figure 11. Sample VR environment of the P11P43P71P82 combination [Source: Author].

4.4. *Verification of the Combination P12P42P71P81*

Validation of the P12P42P71P81 combination was performed on a production process with a modified Z-score of 14.88. This was an existing workstation. Table 13 shows a specific manufacturing process from an automotive company.

Table 13. Case Study P12P42P71P81 [Source: Author].

Technical aspect	Ongoing operation time	Modified Z-score result
Brownfield	20–60 min	14.88
Human Resources (Qualifications)	Continuous production preparation	Suitable for modern visualisation tools
Low	6–60 s	YES
Impact in case of error	Personnel flows (place attachment)	Type of visualisation tool
Medium	Above 80%	VR 2D
Production repeatability	Work execution	Position
Serial production	Parallel manual work up to 50%	Stationary

According to the methodology, it was recommended to use a VR 2D stationary instrument. The visualisation was prepared for a 2D stationary display showing virtual reality (see Figure 12). The placement was made fixed at the workstation. In the demonstration, two selected steps of the manufacturing process can be seen. The visualisation is supplemented with text or a colour scale for some steps for clarity. The green colour symbolizes the correct position or placement of the parts. Red shows the incorrect option.

For better orientation of workers in the space, the solution includes a virtual window of the worker's position at the workplace. The window is located in the upper left corner of the demonstration.

This was followed by the deployment of the designed variant in a real production process and testing with the help of production workers and production managers. As a result, the suitability of the VR 2D stationary solution for this particular combination of parameters was demonstrated.

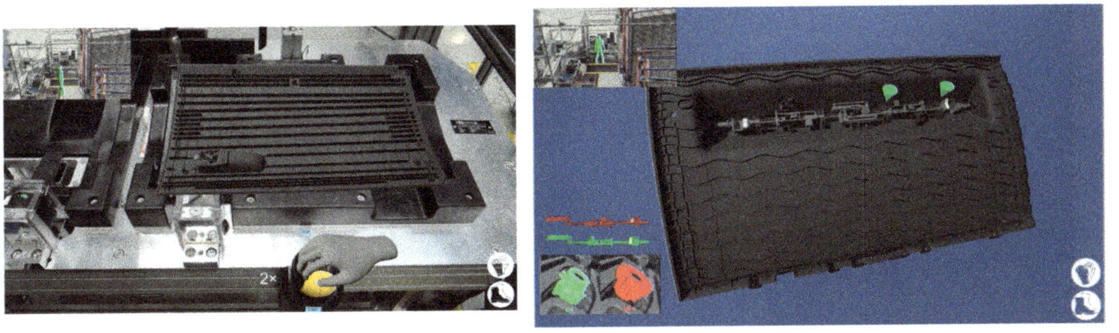

Figure 12. Example of the implementation of the P12P42P71P81 combination [Source: Author].

4.5. Verification of the Combination P12P42P71P82

Validation of the P12P42P71P82 combination was performed on a manufacturing process with a modified Z-score of 16.00. In Table 14 can be seen the parameters of the manufacturing process in a company focused on the production of components for the automotive industry.

Table 14. Case Study P12P42P71P82 [Source: Author].

Technical aspect	Ongoing operation time	Modified Z-score result
Brownfield	1–20 min	16.00
Human Resources (Qualifications)	Continuous production preparation	Suitable for modern visualisation tools
Low	6–60 s	YES
Impact in case of error	Personnel flows (place attachment)	Type of visualisation tool
Medium	Over 80%	VR 2D
Production repeatability	Work execution	Position
Serial production	Parallel manual work over 50%	Stationary

The type of VR 2D stationary visualization tool was selected based on the proposed methodology. The display was on a stationary 2D monitor suspended at the workstation. The selected visualization tool allows the selection of the specific type of product to be produced and it is possible to switch between the types. The steps are aligned in virtual reality with the real working time to avoid inaccurate display. It is possible to scroll between the steps of the visualization by pressing the arrows in the bottom corners of the touch screen as is shown in Figure 13.

The designed visualization tool was then deployed and tested in a real industrial enterprise environment. The VR 2D stationary solution solved the long-term complications of the enterprise within the process. This means that the type and position of the visualization tool selected by the proposed methodology is proven to be very effective.

4.6. Verification of the Combination P12P42P72P81

The P12P42P72P81 combination was validated on a production process with a modified Z-score of 14.31. This is an existing workstation which has a high emphasis on error prevention and the personnel flow around the workstation, as can be seen in Table 15.

Using the proposed methodology, the suitability of deploying modern visualization tools was assessed and a 2D portable VR type was selected, as can be seen in the actual display was performed on a 2D portable industrial display, shown in Figure 14.

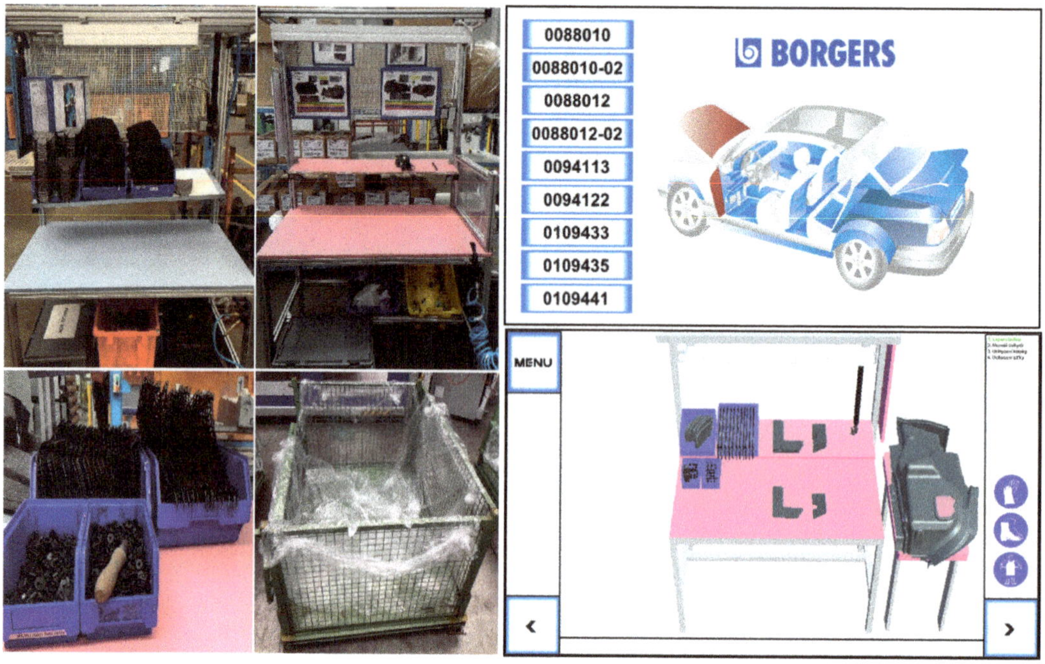

Figure 13. Example of the implementation of the combination P12P42P71P82 [Source: Author].

Table 15. Case Study P12P42P72P81 [Source: Author].

Technical aspect	Ongoing operation time	Modified Z-score result
Brownfield	20–60 min	14.31
Human Resources (Qualifications)	Continuous production preparation	Suitable for modern visualisation tools
Low	60 s and more	YES
Impact in case of error	Personnel flows (place attachment)	Type of visualisation tool
High	Up to 80%	VR 2D
Production repeatability	Work execution	Position
Serial production	Parallel manual work up to 50%	Portable

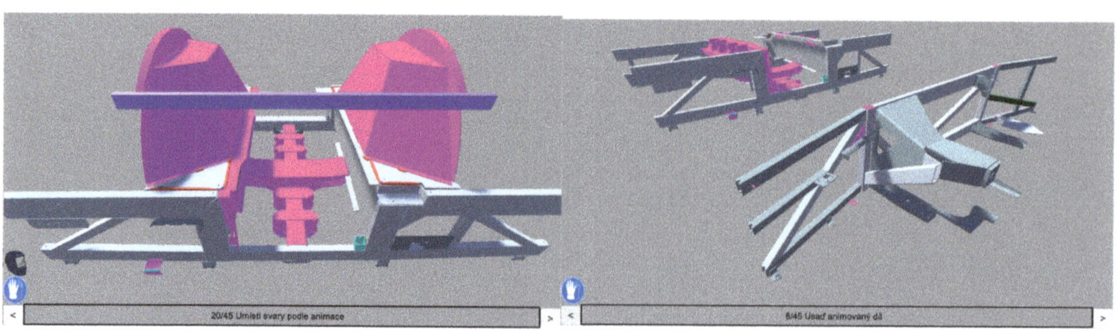

Figure 14. Example of P12P42P72P81 combination implementation [Source: Author].

The selected process has a high personnel flow, which makes it impossible to use stationary solutions. Therefore, the option of a special industrial tablet was chosen. This tablet must also be able to be operated while wearing gloves. At the same time, the production process is carried out in a dusty environment, and it was necessary to choose a device that is suitable for the environment. It is controlled by simply switching between steps using arrow buttons in the bottom corners of the screen. For easy orientation in the process, a bottom bar with a description and a numerical designation of the visualization step is used.

This solution enables the movement of the worker together with the visualization tool to ensure its constant availability. This option has been assessed as satisfactory.

4.7. Verification of the Combination P12P42P72P82

The P12P42P72P82 combination was validated on a production process with a modified Z-score of 15.41. The site already exists. Here the key needs are for frequent work using both hands simultaneously and the large distance of personnel flows, as can be seen in Table 16.

Table 16. Case Study P12P42P72P82 [Source: Author].

Technical aspect	Ongoing operation time	Modified Z-score result
Brownfield	1–20 min	15.41
Human Resources (Qualifications)	Continuous production preparation	Suitable for modern visualisation tools
Medium	6–60 s	YES
Impact in case of error	Personnel flows (place attachment)	Type of visualisation tool
High	Up to 80%	AR
Production repeatibility	Work execution	Position
Serial production	Parallel manual work over 50%	Portable

In Figure 15 it can be seen that a portable AR headset has been selected for the given combination by the proposed methodology. It allows working with both hands simultaneously and complements the real world with virtual elements using a semi-transparent display. The virtual elements are displayed directly into the real world and help navigate the worker while performing the work. The navigation elements can take the form of various shapes or text labels. This combination can be used to clearly identify problem areas while the work is being performed. This solution also allows for the acceptance of demanding personnel flows along with parallel manual work. A key feature of the chosen AR headset is the ability to control it using gestures, so there is no need for additional hand controllers.

Figure 16 can be seen the recognition of production equipment by means of special markers that allow the equipment or production process to be uniquely identified. After loading the device after recognition, we gesture to the process to be executed. The next steps are carried out in a similar way as in the previous combinations, by means of arrow buttons that allow orientation in the process and its sub-steps.

The selected visualization tool was then deployed and tested in the real environment of an industrial enterprise. Production operators, machine operators and adjusters participated in the testing. The portable AR headset was judged by the expert group and company representatives to be suitable for this production process. However, an important finding is that AR technology does not yet allow workers to use the device for the entire work shift. The recommended AR time for this case was set at a maximum of 2 h per day.

4.8. Verification of the Combination P12P43P71P82

To verify the combination P12P43P71P82, a process corresponding to the characteristics of this variant was selected. However, this is not a manufacturing company, but a logistics centre where goods are packed into boxes in combination with work with an information system. This can be considered sufficient for the purpose of validating the methodology and assessing this sub-combination. The value of the resulting Z-score meets the conditions, as can be seen in Table 17.

For this combination, the methodology selected a stationary video-guide as a suitable option. An important prerequisite is an almost constant presence at the workplace in order to be able to capture the individual steps by video recording. At the same time, it is necessary to ensure good visibility throughout the steps to be captured on video. This type of visualisation tool also assumes a small variation in products (in this case packaging methods). Recording a large number of products is not advisable due to the need to take a lot of footage and the complexity of making adjustments if any of the steps are changed.

Figure 15. Example of a visualization tool in AR [Source: Author].

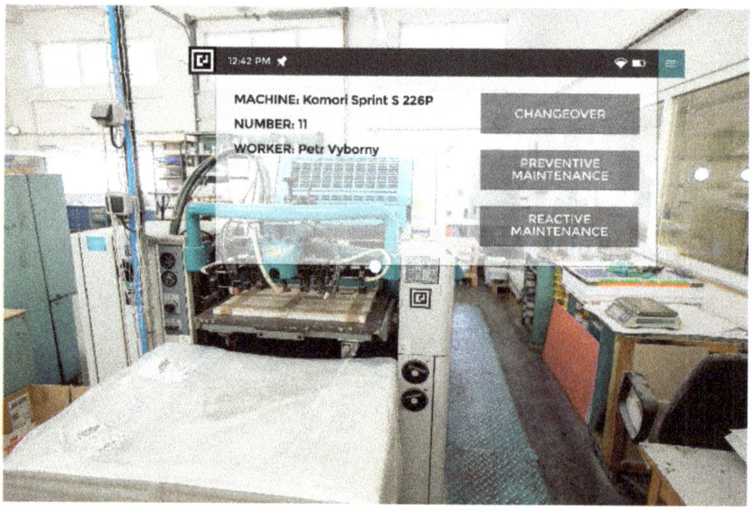

Figure 16. Example of P12P42P72P82 combination implementation [Source: Author].

The verification took place during real operation with representatives of the company. The solution was evaluated by the expert group as satisfactory. However, prerequisites for implementation are a good view of the workplace, the location of the individual steps and the worker's hands.

Table 17. Case Study P12P43P71P82 [Source: Author].

Technical aspect	Ongoing operation time	Modified Z-score result
Brownfield	1–20 min	14.80
Human Resources (Qualifications)	Continuous production preparation	Suitable for modern visualisation tools
Low	6–60 s	YES
Impact in case of error	Personnel flows (place attachment)	Type of visualisation tool
High	Over 80%	Video
Production repeatability	Work execution	Position
Bulk	Parallel manual work over 50%	Stationary

4.9. Verification of the Combination P12P43P72P81

The P12P43P72P81 combination was validated on a production process with a modified Z-score of 14.80. This is an existing workplace, with a high emphasis on error prevention and the personnel flow around the workplace. This is reflected in the parameters in Table 18.

For this combination, the methodology selected a portable VR 2D tool. In Figure 17 an example implementation can be seen. The display was via a portable 2D industrial monitor. This is due to the high personnel flow with a relatively large distance of worker movement.

The example shows the output of a visualization tool focused on a semi-automated manufacturing process. The worker operates the machine and prepares the necessary material at the input, including material handling at the output.

As seen in Figure 18, the VR 2D environment is made up of several auxiliary elements to make working with the selected tool as simple and efficient as possible. In the right part there is a window with a breakdown of the steps, where the step of the current process is also marked. The steps are again controlled by the arrows in the bottom corners. Each step has an additional text description. For some of the steps, the necessary protective equipment is also added to prevent possible work accidents. The result meets expectations and allows the worker to move freely around the workplace. The industrial display is thus available to the worker at all times. This also helps eliminate the risk of errors or incorrect sequencing of process steps.

Table 18. Case Study P12P43P72P81 [Source: Author].

Technical aspect	Ongoing operation time	Modified Z-score result
Brownfield	1–20 min	14.80
Human Resources (Qualifications)	Continuous production preparation	Suitable for modern visualisation tools
Low	6–60 s	YES
Impact in case of error	Personnel flows (place attachment)	Type of visualisation tool
High	Up to 80%	VR 2D
Production repeatability	Work execution	Position
Mass production	Parallel manual work up to 50%	Portable

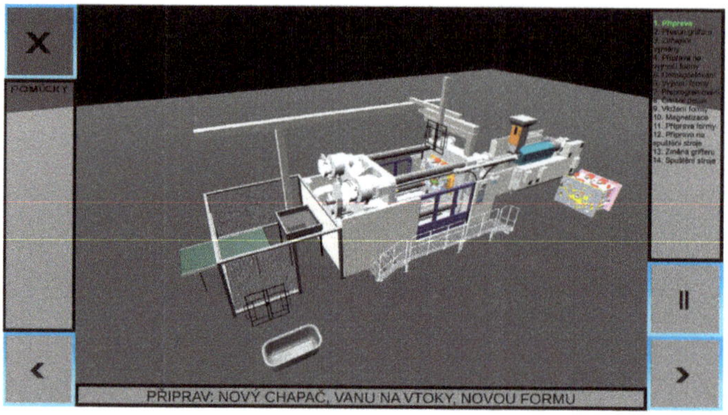

Figure 17. Example of the P12P43P72P81 combination [Source: Author].

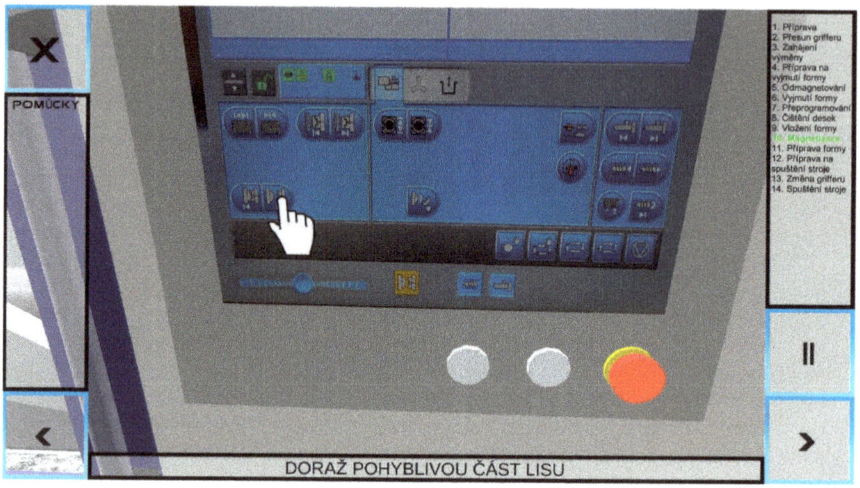

Figure 18. Two-dimensional VR environment [Source: Author].

4.10. Verification of the Combination P12P43P72P82

The last verified combination is P12P43P72P82. This is an example of mass production, which is characterized by low product variation and very fast production cycles. The validation took place at the company's existing automotive site. According to Table 19 the modified Z-score for this process is 14.80.

Table 19. Case Study P12P43P72P82 [Source: Author].

Technical aspect	Ongoing operation time	Modified Z-score result
Brownfield	1–20 min	14.80
Human Resources (Qualifications)	Continuous production preparation	Suitable for modern visualisation tools
Low	6–60 s	YES
Impact in case of error	Personnel flows (place attachment)	Type of visualisation tool
High	Up to 80%	VR 2D
Production repeatibility	Work execution	Position
Mass production	Parallel manual work over 50%	Stationary

According to the methodology, a stationary VR 2D visualization tool is used. The implementation was carried out by creating a virtual reality application that is displayed on a screen directly at the workplace. All the elements necessary for the production process were modelled to correspond 1:1 to the real situation at the workplace.

In Figure 19 it can be seen that the individual steps are accompanied not only by text descriptions but also by graphic elements that indicate the correct execution of the step or the risk of error. Orientation in the process is provided in the right column where all steps are displayed and the current step being performed is highlighted in green. To increase the efficiency of the designed variant, a connection between the visualization tool and the production equipment was created. This connection is provided by a Programmable Logic Control (PLC) unit that receives inputs from the production machine and transmits information to the visualization tool about the step of the production process. This communication makes it possible to eliminate the interaction of the worker with the visualization tool. Thus, the worker is oriented only by looking at the visualization tool, which displays the procedure for the step they are currently in.

Figure 19. Two-dimensional VR environments (Work with the part according to the animation.) [Source: Author].

The VR 2D solution was therefore connected directly to the production equipment (see Figure 20) to synchronise the production steps and the visualisation tool. Testing was carried out under real production conditions. The suitability and functionality of the proposed type of visualisation tool for a given combination of parameters was demonstrated with company representatives and the expert group.

Figure 20. Example of P12P43P72P82 combination implementation [Source: Author].

5. Implementation and Results

As already mentioned, verification was carried out in ten case studies. These case studies were carried out in parallel in cooperation between the academic and commercial spheres (University of West Bohemia in Pilsen and a start-up company that emerged from the university). The implementation time of each project was from 3 to 8 months from design through implementation. Each sub-project was quite different and involved a new innovative approach in implementation. Each project also involved many meetings at the managerial and executive level. For protection reasons, it is not possible to name specific companies at this point, nor to describe each implementation in detail.

The aim of the validation was to evaluate the functionality of the proposed methodology. In order to make the methodology complete, it was necessary to confirm the outputs of the methodology in real applications in industrial practice. Thus, the methodology must be satisfactory in terms of the selected characteristics of the production processes, the proposed multicriteria function and the matrix of the combinations of type and method of use. All this cannot be demonstrated by only one deployment. It was therefore necessary to validate at least three different combinations of type P11 and all possible combinations of type P12.

For the verification of the P11 combinations, three specific process variants were selected as far as possible. The condition for inclusion in the verification part was to meet a minimum Z-score of 14.

Individual combinations and outputs P11:

- P11P42P71P82-VR 3D, field tested, verified;
- P11P42P72P81-VR 3D, field tested, verified;
- P11P43P71P82-VR 3D, field tested, verified.

Validation of the P12 combinations was carried out by testing all combinations in industrial practice. In total, there were eight possible combinations. Of these, seven combinations were successfully implemented and verified. One combination was not implemented due to not finding a suitable production process.

Individual combinations and outputs P12:

- P12P42P71P81-VR 2D stationary, field tested, verified;
- P12P42P71P82-VR 2D stationary, field tested, verified;
- P12P42P72P81-VR 2D portable, field tested, verified;
- P12P42P72P82-AR portable, field tested, verified;
- P12P43P71P81-video guide, not tested, not verified;
- P12P43P71P82-video tutorial, tested in practice, verified;
- P12P43P72P81-VR 2D portable, field tested, verified;
- P12P43P72P82-VR 2D stationary, field tested, verified.

The creation of each individual case study and the facilitation of this creation through the methodology has been made possible by a rare synergy of commercial and academic sector needs. The execution of the ten combinations is absolutely unique not only because of the huge amount of human and material resources used. Making a rigor and objective quantitative evaluation is difficult because many factors enter into it. Rather, it seems more appropriate to use qualitative evaluation based on descriptive research methodology [67]. Based on this, production managers were interviewed six months after the deployment of each application. On the basis of these interviews, in all cases a positive benefit was expressed from a subjective point of view, even in nine cases (out of ten) there was an objective increase in productivity and in all cases a reduction in scrap rates.

From the above validations of both P11 and P12 based combinations, it was concluded that the methodology seems to be functional and has been successfully used by 10 implementations in real production processes. The time requirement for the implementation of the solution is very high and for this reason 10 such verifications were performed during the 2 years of the research. Although the P12P43P71P81 combination was not verified, the number of tests implemented was sufficient.

Thanks to the developed software support, it is also possible to use the proposed methodology without detailed knowledge of the individual components and algorithms contained in the methodology. This is also a key factor for the future application of this methodology in practice. For further applications, it is mainly assumed that the values of individual characteristics of the production process are correctly defined and that they are sufficiently known.

6. Discussion

An important part of this work is the verification of the proposed methodology in practice. A method of evaluation of the methodology was proposed by implementing 10 projects in real conditions of industrial enterprises. The aim of the projects was to verify the functionality and contribution of the methodology. To perform this, it was necessary not only to evaluate each project and select a suitable visualization tool, but also to build a complete solution that could be tested in real operation. Each sub-project was therefore thoroughly prepared and analysed. This was followed by the creation of models or recordings. In some cases, a completely new virtual environment with many interactive elements was created. The improvement of process efficiency by using modern visualisation elements compared with paper-based instructions has already been confirmed in previous research by the authors [68].

Division of the methodology development process into phases

The process of developing a methodology consists of four phases: preparation, decision-making, evaluation, and implementation. In the first phase, the needs and objectives are analysed in detail and a basic framework is established. This is followed by decisions on the key elements and structure of the methodology, including the selection of appropriate methods and tools. The third phase involves the evaluation of the methodology against the established criteria, with emphasis on identifying gaps and opportunities for improvement. The last phase is dedicated to the implementation of the methodology, including implementation, staff training and performance monitoring. It is crucial to note that these phases are not isolated but interact, and harmonisation between them is essential for a successful outcome of the whole process.

The nature of manufacturing processes and visualization tools relationship

Relationship between the nature of manufacturing processes and visualization tools

The existence of a dependency between the choice of visualisation tools for the creation of instructions within the organisational preparation of production and the nature of the production processes was assured. A Likert scale was used to find the relationship between the manufacturing process and modern visualization tools, which was modified by the author to fit the focus of the research. The individual characteristics of the manufacturing processes were then examined to determine the impact on the choice of visualization tools and the strength of the relationship. The identification of the key characteristics is the focus of the same chapter in which the research and statistical evaluation of the modified Likert scale output is conducted. The relevant characteristics of manufacturing processes with a strong link to the choice of visualisation tools are retrieved Figure 3 then evaluated and the results of the research can be seen.

It was found that it is possible to assess the suitability of using modern visualization tools for a given process according to its characteristics. The characteristics of the production processes that have a significant influence on the choice of visualization tools were divided into three groups according to whether they influence the choice of type, method of use of the visualization tool or both at the same time. In particular, the characteristics that are important for testing this hypothesis are those belonging to groups A and B in Table 2.

Criterium function and parameters defining

The next step was the design of a multicriteria function based on Altman's model, which was revised by the author in evaluation phase. Based on this model, the design of the multicriteria function resulted in a modified Z-score. The function accounts for the characteristics of groups A and B, which are represented in the calculation by their strength of association and the value of each parameter. Based on the resulting modified Z-score, the production processes are divided into three main zones. This division is based on the result of the simulations and the reliability test, which can be seen in Figure 5. These are the zones of conventional tools, the grey zones and the zones of using modern visualization tools. It is the last mentioned zone, defined by $14 \leq Z_{MOD}$, that represents the zone where it is appropriate and efficient to use modern visualization tools. The methodology for selecting a visualization tool based on suitability has not yet been sufficiently developed in previous research [69]. This paper shows that the proposed methodology can be used to select the appropriate type of visualization tool for a specific manufacturing process. The result function proposed in section evaluation phase characterized as a modified Z-score is an indicator of the suitability of deploying modern visualization tools. Furthermore, if the production process falls within the zone of use of modern visualization tools, it is necessary to decide which type of visualization tools will be selected and how they will be used. The inappropriateness of a combination of paper and digital manuals has already been demonstrated [9]. The characteristics of manufacturing processes that influence the selection of a modern visualization tool belong to groups A and C.

Developing a methodology tool and verification

In section implementation phase it can be seen that combining these characteristics results in 24 possible combinations. The different combinations are further statistically evaluated, and the result can be seen in Tables 7 and 8 The results were also examined in terms of meeting the rules of relevance of the data obtained. The output is a matrix combining the characteristics of the production processes and a suitable visualization tool. This is the last step of the proposed methodology, which has been completed for ease of use in section creation of the software support and then in validation of the method verified by implementing the individual combinations in industrial practice.

The proposed methodology seems to work within pre-set limits that indicate where the use of this methodology is appropriate. However, the methodology has not been designed and validated outside the pre-set limits. It is therefore uncertain how the methodology will behave in the event of exposure outside the defined limit constraints. Therefore, the future direction of this research should be to extend the limit constraints and verify the properties of the proposed methodology in this area. The survey was carried out on a sufficient number of respondents, but all the companies involved were based in the Czech Republic. It would therefore be useful to see if a similar situation applies in other countries. This would allow the attitude of enterprises abroad to be assessed and, if necessary, software support to be adapted for the purposes of a particular country. Another benefit could be to test the methodology and combinations on a larger number of implemented projects. This would allow validation on a larger number of implementations or possible adjustments resulting from practical experience. The validation also lacks the implementation of the combination P12P43P71P81, which could not be secured. Validation of this combination would then conclude the validation of all possible combinations mentioned in section validation of the method.

7. Conclusions

The main theoretical contribution of this paper is the development of a methodology for assessing the suitability of deployment and use of modern visualization tools. This is related to the identification of the current state of knowledge within the topics related to the product life cycle and technical preparation of production. This is followed by a section focusing on the use of modern visualization tools and linking them to the characterization of production process parameters. These tools are tutorials in the form of video, augmented reality, and virtual reality. The characteristics of the production processes that have a significant influence on the choice of visualization tools were divided into three groups according to whether they influence the choice of type, method of use of the visualization tool or both at the same time. A multicriteria function based on the modified Altman model was then proposed, resulting in a modified Z-score. The score is essential for grouping manufacturing processes according to their suitability for deploying modern visualization tools. Specifically, these are the conventional tool zones, the grey zones, and the modern visualization tool utilization zones. Another proposed tool is the basis for deciding what type of visualization tools will be chosen and how they will be used. The assumption is that the production process falls within the zone of use of modern visualization tools, which is defined by a multicriteria function from the characteristics of the production processes that influence the selection of a modern visualization tool. All conceivable combinations of the investigated parameters were compiled. The combinations were then statistically evaluated, and the research resulted in a matrix of combinations of production process characteristics and appropriate use of visualization tools. This step completed the methodology. The important part is the verification of the proposed methodology in practice. Here, a way of evaluating the methodology was proposed by implementing 10 projects in real conditions of industrial enterprises. The aim of the projects was to verify the functionality and contribution of the methodology. To perform this, it was necessary not only to evaluate each project and select a suitable visualization tool, but also to build a complete solution that could be tested in real operation. Each sub-project was therefore thoroughly prepared and analysed. This was followed by the creation of models or recordings. In some cases, a completely new virtual environment with many interactive elements was created. Everything was then communicated not only within the expert group, but also with the managers of the individual companies and the workers who then used the selected visualisation tool. This validation contributed to the verification and finalisation of the methodology. The result is a methodology that facilitates the choice of an appropriate visualization tool in the organizational preparation of production and serves as an aid in the decision-making of manufacturing companies for the deployment of modern visualization tools. Crucially, the effectiveness of the proposed methodology is not confined to theoretical constructs. The research goes a step further by introducing a robust validation process. Ten real industrial projects serve as the testing ground, allowing for the evaluation of the methodology's functionality and contribution in practical, operational settings. Each project undergoes thorough preparation, analysis, and, in some cases, the creation of entirely new virtual environments with interactive elements. The methodology's application is not limited to academic circles; it extends to interactions with managers and workers within industrial enterprises. This real-world validation contributes significantly to verifying and finalizing the methodology, ensuring its relevance and applicability in diverse industrial contexts. In conclusion, the result of this research is a methodology that transcends theoretical frameworks. It stands as a practical and valuable decision-making aid for manufacturing companies in the organizational preparation for production. By seamlessly integrating theoretical contributions and real-world validation, this methodology represents a holistic approach to addressing the challenges of deploying modern visualization tools in industrial settings. In future work, it would be advisable to focus on extending the multicriteria function based on the modified Altman model and possibly use and validate the proposed methodology in foreign markets outside the Czech Republic. Furthermore, it would be interesting to extend the application of the methodology to different industries or sectors, to assess its effectiveness in different contexts and to identify possible sector specificities.

Author Contributions: Conceptualization, K.N. and P.H. methodology, K.N. validation, K.N., J.K. and M.D.; formal analysis, K.N. and D.K. investigation, K.N. and P.H.; resources, J.K., M.D., M.B. and D.K. writing—original draft preparation, K.N. and P.H.; writing—review and editing, K.N., M.K. and P.H.; visualization, J.K., M.D., M.B. and D.K.; supervision, P.H. and M.Š.; project administration, P.H.; and funding acquisition, M.Š. All authors have read and agreed to the published version of the manuscript.

Funding: This research was funded by the Internal Science Foundation of the University of West Bohemia under Grant SGS-2021-028 'Developmental and Training Tools for the Interaction of Man and the Cyber–Physical Production System'.

Data Availability Statement: The datasets and graphs generated during the current study are available in the Google Drive repository: https://drive.google.com/drive/folders/1XpJhgRJUHtwEPMZBSTg5olQAISR9RtrF?usp=sharing. Accessed on 6 September 2023.

Conflicts of Interest: The authors declare no conflict of interest.

References

1. Saeed, H.M.M.; Ali, A.; Hameed, D.M. The Role Of Product Lifecycle Cost Technology And Cleaner Production Strategy In Reducing Costs. *J. Posit. Sch. Psychol.* **2022**, 1125–1139.
2. Regassa Hunde, B.; Debebe Woldeyohannes, A. Future Prospects of Computer-Aided Design (CAD)–A Review from the Perspective of Artificial Intelligence (AI), Extended Reality, and 3D Printing. *Results Eng.* **2022**, *14*, 100478. [CrossRef]
3. Šulák, M.; Vacík, E. *Strategické Řízení v Podnicích a Projektech*, 1st ed; Vysoká škola finanční a správní: Praha, Czech Republic, 2005; ISBN 80-86754-35-9.
4. Rosenau, M.D. *Successful Project Management*, 3rd ed.; John Wiley and Sons: New York, NY, USA, 1998; ISBN 0-471-29304-0.
5. Fotr, J.; Souček, I. *Investiční Rozhodování a Řízení Projektů*, 1st ed.; Grada Publishing: Praha, Czech Republic, 2011; ISBN 978-80-247-3293-0.
6. Fernandez, I.; Casla, P.; Esnaola, I.; Parigot, L.; Marguglio, A.; Johnson, T. Towards Adaptive, Interactive, Assistive and Collaborative Assembly Workplaces Through Semantic Technologies. In *Proceedings of the Enterprise Interoperability IX.*; Archimède, B., Ducq, Y., Young, B., Karray, H., Eds.; Springer International Publishing: Cham, Switzerland, 2023; pp. 39–49.
7. Dragomiroiu, R.; Hurloiu, I.-I.; Mihai, G. Induction Staff Training. *Procedia Econ. Financ.* **2014**, *16*, 368–373. [CrossRef]
8. Eversberg, L.; Lambrecht, J. Evaluating Digital Work Instructions with Augmented Reality versus Paper-Based Documents for Manual, Object-Specific Repair Tasks in a Case Study with Experienced Workers 2023. *arXiv* **2023**, arXiv:2301.07570.
9. Letmathe, P.; Rößler, M. Should Firms Use Digital Work Instructions?—Individual Learning in an Agile Manufacturing Setting. *J. Oper. Manag.* **2022**, *68*, 94–109. [CrossRef]
10. Ross, J.; Schulz, R. Can Computer-Aided Instruction Accommodate All Learners Equally? *Br. J. Educ. Technol.* **1999**, *30*, 5–24. [CrossRef]
11. Busck, N.; Svensson, F. Assembly Instructions for Swedish Manufacturing Industry of the Future. Master's Thesis, Chalmers University of Technology, Gothenburg, Sweden, 2017.
12. Nikolenko, A.; Sehr, P.; Hinrichsen, S.; Bendzioch, S. Digital Assembly Assistance Systems–A Case Study. In Proceedings of the Advances in Human Factors and Systems Interaction; Nunes, I.L., Ed.; Springer International Publishing: Cham, Switzerland, 2020; pp. 24–33.
13. Chan, Y.M. Video Instructions as Support for beyond Classroom Learning. *Procedia–Soc. Behav. Sci.* **2010**, *9*, 1313–1318. [CrossRef]
14. Alemdag, E. A Scoping Review of the Literature on Embodied Instructional Videos. *Res. Pract. Technol. Enhanc. Learn.* **2023**, *18*, 29. [CrossRef]
15. Teach Me How! Interactive Assembly Instructions Using Demonstration and In-Situ Projection. Available online: https://www.researchgate.net/publication/321019188_Teach_Me_How_Interactive_Assembly_Instructions_Using_Demonstration_and_In-Situ_Projection (accessed on 6 December 2023).
16. van der Meij, H.; van der Meij, J. A Comparison of Paper-Based and Video Tutorials for Software Learning. *Comput. Educ.* **2014**, *78*, 150–159. [CrossRef]
17. Reviews in Instructional Video. Available online: https://www.researchgate.net/publication/318197434_Reviews_in_instructional_video (accessed on 6 December 2023).
18. Carmigniani, J.; Furht, B. Augmented Reality: An Overview. In *Handbook of Augmented Reality*; Furht, B., Ed.; Springer: New York, NY, USA, 2011; pp. 3–46. ISBN 978-1-4614-0064-6.
19. Wang, Z.; Bai, X.; Zhang, S.; Billinghurst, M.; He, W.; Wang, P.; Lan, W.; Min, H.; Chen, Y. A Comprehensive Review of Augmented Reality-Based Instruction in Manual Assembly, Training and Repair. *Robot. Comput.-Integr. Manuf.* **2022**, *78*, 102407. [CrossRef]
20. Deshpande, A.; Kim, I. The Effects of Augmented Reality on Improving Spatial Problem Solving for Object Assembly. *Adv. Eng. Inform.* **2018**, *38*, 760–775. [CrossRef]
21. Yan, W. Augmented Reality Instructions for Construction Toys Enabled by Accurate Model Registration and Realistic Object/Hand Occlusions. *Virtual Real.* **2022**, *26*, 465–478. [CrossRef] [PubMed]
22. Mourtzis, D.; Xanthi, F.; Zogopoulos, V. An Adaptive Framework for Augmented Reality Instructions Considering Workforce Skill. *Procedia CIRP* **2019**, *81*, 363–368. [CrossRef]
23. Mourtzis, D.; Zogopoulos, V.; Katagis, I.; Lagios, P. Augmented Reality Based Visualization of CAM Instructions towards Industry 4.0 Paradigm: A CNC Bending Machine Case Study. *Procedia CIRP* **2018**, *70*, 368–373. [CrossRef]
24. Jensen, L.; Konradsen, F. A Review of the Use of Virtual Reality Head-Mounted Displays in Education and Training. *Educ. Inf. Technol.* **2018**, *23*, 1515–1529. [CrossRef]
25. Seo, J.; Avellino, I.; Rajasagi, D.P.; Komlodi, A.; Mentis, H.M. HoloMentor: Enabling Remote Instruction through Augmented Reality Mobile Views. *Proc. ACM Hum.-Comput. Interact.* **2022**, *7*, 1–29. [CrossRef]
26. Marques, B.; Ferreira, C.; Silva, S.; Dias, P.; Santos, B. Is Social Presence (Alone) a General Predictor for Good Remote Collaboration? Comparing Video and Augmented Reality Guidance in Maintenance Procedures. *Virtual Real.* **2023**, *27*, 1783–1796. [CrossRef]

27. Palmarini, R.; Erkoyuncu, J.A.; Roy, R.; Torabmostaedi, H. A Systematic Review of Augmented Reality Applications in Maintenance. *Robot. Comput.-Integr. Manuf.* **2018**, *49*, 215–228. [CrossRef]
28. Oliveira, R.; Farinha, J.; Raposo, H.; Pires, J.N. Augmented Reality and the Future of Maintenance. In Proceedings of the Maintenance Performance Measurement and Management (MPMM), Coimbra, Portugal, 4–5 September 2014; ISBN 978-972-8954-42-0.
29. Konstantinidis, F.K.; Kansizoglou, I.; Santavas, N.; Mouroutsos, S.G.; Gasteratos, A. MARMA: A Mobile Augmented Reality Maintenance Assistant for Fast-Track Repair Procedures in the Context of Industry 4.0. *Machines* **2020**, *8*, 88. [CrossRef]
30. Henderson, S.; Feiner, S. Exploring the Benefits of Augmented Reality Documentation for Maintenance and Repair. *IEEE Trans. Vis. Comput. Graph.* **2011**, *17*, 1355–1368. [CrossRef]
31. Manuri, F.; Pizzigalli, A.; Sanna, A. A State Validation System for Augmented Reality Based Maintenance Procedures. *Appl. Sci.* **2019**, *9*, 2115. [CrossRef]
32. Nanjappan, V.; Uunila, A.; Vaulanen, J.; Välimaa, J.; Georgiev, G.V. Effects of Immersive Virtual Reality in Enhancing Creativity. *Proc. Des. Soc.* **2023**, *3*, 1585–1594. [CrossRef]
33. Ulmer, J.; Braun, S.; Cheng, C.-T.; Dowey, S.; Wollert, J. Gamified Virtual Reality Training Environment for the Manufacturing Industry. In Proceedings of the 2020 19th International Conference on Mechatronics–Mechatronika (ME), Prague, Czech Republic, 2–4 December 2020; pp. 1–6.
34. Radhakrishnan, U.; Chinello, F.; Koumaditis, K. Immersive Virtual Reality Training: Three Cases from the Danish Industry. In Proceedings of the 2021 IEEE Conference on Virtual Reality and 3D User Interfaces Abstracts and Workshops (VRW), Lisbon, Portugal, 27 March–3 April 2021; pp. 1–5.
35. Balzerkiewitz, H.P.; Schade, N.; Stechert, C. Evaluation of the Learning Effect of VR on Engineering Education–Case Study in Machine Elements. In Proceedings of the 2022 IEEE International Conference on Industrial Engineering and Engineering Management (IEEM), Kuala Lumpur, Malaysia, 7–10 December 2022; pp. 1252–1256.
36. Merchant, Z.; Goetz, E.T.; Cifuentes, L.; Keeney-Kennicutt, W.; Davis, T.J. Effectiveness of Virtual Reality-Based Instruction on Students' Learning Outcomes in K-12 and Higher Education: A Meta-Analysis. *Comput. Educ.* **2014**, *70*, 29–40. [CrossRef]
37. Patle, D.S.; Manca, D.; Nazir, S.; Sharma, S. Operator Training Simulators in Virtual Reality Environment for Process Operators: A Review. *Virtual Real.* **2019**, *23*, 293–311. [CrossRef]
38. Wolfartsberger, J.; Zimmermann, R.; Obermeier, G.; Niedermayr, D. Analyzing the Potential of Virtual Reality-Supported Training for Industrial Assembly Tasks. *Comput. Ind.* **2023**, *147*, 103838. [CrossRef]
39. Rosen, S.; Bricken, W.; Martinez, R.; Laurel, B.; Mitchell, A.R. Determinants of Immersivity in Virtual Reality: Graphics vs. Action. In Proceedings of the 21st Annual Conference on Computer Graphics and Interactive Techniques, Orlando, FL, USA, 24–29 June 1994; Association for Computing Machinery: New York, NY, USA; p. 496.
40. Gupta, S.; Wilcocks, K.; Matava, C.; Wiegelmann, J.; Kaustov, L.; Alam, F. Creating a Successful Virtual Reality-Based Medical Simulation Environment: Tutorial. *JMIR Med. Educ.* **2023**, *9*, e41090. [CrossRef]
41. Dorloh, H.; Li, K.-W.; Khaday, S. Presenting Job Instructions Using an Augmented Reality Device, a Printed Manual, and a Video Display for Assembly and Disassembly Tasks: What Are the Differences? *Appl. Sci.* **2023**, *13*, 2186. [CrossRef]
42. Gors, D.; Put, J.; Vanherle, B.; Witters, M.; Luyten, K. Semi-Automatic Extraction of Digital Work Instructions from CAD Models. *Procedia CIRP* **2021**, *97*, 39–44. [CrossRef]
43. Chowdhury, A.M.; Moon, S. Generating Integrated Bill of Materials Using Mask R-CNN Artificial Intelligence Model. *Autom. Constr.* **2023**, *145*, 104644. [CrossRef]
44. Palmarini, R.; Del Amo, I.F.; Ariansyah, D.; Khan, S.; Erkoyuncu, J.A.; Roy, R. Fast Augmented Reality Authoring: Fast Creation of AR Step-by-Step Procedures for Maintenance Operations. *IEEE Access* **2023**, *11*, 8407–8421. [CrossRef]
45. Neb, A.; Strieg, F. Generation of AR-Enhanced Assembly Instructions Based on Assembly Features. *Procedia CIRP* **2018**, *72*, 1118–1123. [CrossRef]
46. Luki, D.; Todi, V.; Miloševi, M. Model of Modern Technological Production Preparation. *Proc. Manuf. Syst.* **2010**, *5*, 15–22.
47. Lyubomudrov, S.A.; Khrustaleva, I.N.; Tolstoles, A.A.; Maslakov, A.P. Improving the efficiency of technological preparation of single and small batch production based on simulation modeling. *J. Min. Inst.* **2019**, *240*, 669. [CrossRef]
48. Lyrath, F.; Stechert, C.; Ahmed, S.I.-U. Application of Augmented Reality (AR) in the Laboratory for Experimental Physics. *Procedia CIRP* **2023**, *119*, 170–175. [CrossRef]
49. Raska, P.; Ulrych, Z.; Malaga, M. Data Reduction of Digital Twin Simulation Experiments Using Different Optimisation Methods. *Appl. Sci.* **2021**, *11*, 7315. [CrossRef]
50. Lasi, H.; Fettke, P.; Kemper, H.-G.; Feld, T.; Hoffmann, M. Industry 4.0. *Bus. Inf. Syst. Eng.* **2014**, *6*, 239–242. [CrossRef]
51. Akdere, M.; Jiang, Y.; Lobo, F.D. Evaluation and Assessment of Virtual Reality-Based Simulated Training: Exploring the Human-Technology Frontier. *Eur. J. Train. Dev.* **2022**, *46*, 434–449. [CrossRef]
52. Chen, Q.; Shi, H.; Jiang, L. Comparative Study on QMS at Pre-Production Stage in China and UK Baby Dairy Product Industries. *Open J. Soc. Sci.* **2022**, *10*, 470–487. [CrossRef]
53. Park, K.-B.; Lee, J.Y. Comparative Study on the Interface and Interaction for Manipulating 3D Virtual Objects in a Virtual Reality Environment. *Korean J. Comput. Des. Eng.* **2016**, *21*, 20–30. [CrossRef]
54. Nayernia, H.; Bahemia, H.; Papagiannidis, S. A Systematic Review of the Implementation of Industry 4.0 from the Organisational Perspective. *Int. J. Prod. Res.* **2022**, *60*, 4365–4396. [CrossRef]

55. Florescu, A.; Barabas, S. Development Trends of Production Systems through the Integration of Lean Management and Industry 4.0. *Appl. Sci.* **2022**, *12*, 4885. [CrossRef]
56. Hilken, T.; Chylinski, M.; Keeling, D.I.; Heller, J.; de Ruyter, K.; Mahr, D. How to Strategically Choose or Combine Augmented and Virtual Reality for Improved Online Experiential Retailing. *Psychol. Mark.* **2022**, *39*, 495–507. [CrossRef]
57. Rusch, M.; Schöggl, J.-P.; Baumgartner, R.J. Application of Digital Technologies for Sustainable Product Management in a Circular Economy: A Review. *Bus. Strategy Environ.* **2023**, *32*, 1159–1174. [CrossRef]
58. Joshi, A.; Kale, S.; Chandel, S.; Pal, D. Likert Scale: Explored and Explained. *Br. J. Appl. Sci. Technol.* **2015**, *7*, 396–403. [CrossRef]
59. Hayes, N. *Základy Sociální Psychologie*, 1st ed.; Portál: Praha, Czech Republic, 1998; ISBN 80-7178-198-3.
60. Heřman, J. *Řízení Výroby*, 1st ed.; Melandrium: Slaný, Czech Republic, 2001; ISBN 80-86175-15-4.
61. Keřkovský, M.; Valsa, O. *Moderní Přístupy k Řízení Výroby*, 3rd ed.; C.H. Beck: Praha, Czech Republic, 2012; ISBN 978-80-7179-319-9.
62. Řepa, V. *Podnikové Procesy*, 2nd ed.; Grada Publishing: Praha, Czech Republic, 2007; ISBN 978-80-247-2252-8.
63. Líbal, V. *Organizace a Řízení Výroby*, 7th ed.; SNTL: Praha, Czech Republic, 1989; ISBN 80-03-00050-5.
64. Altman, E.I. *Bankruptcy, Credit Risk, and High Yield Junk Bonds*, 1st ed.; Blackwell: Malden, MA, USA, 2002; ISBN 0-631-22563-3.
65. Cotton, J.L.; Tuttle, J.M. Employee Turnover: A Meta-Analysis and Review with Implications for Research. *Acad. Manag. Rev.* **1986**, *11*, 55–70. [CrossRef]
66. Autor, D. The Labor Market Impacts of Technological Change: From Unbridled Enthusiasm to Qualified Optimism to Vast Uncertainty. *SSRN Electron. J.* **2022**. [CrossRef]
67. Koh, E.T.; Owen, W.L. Descriptive Research and Qualitative Research. In *Introduction to Nutrition and Health Research*; Koh, E.T., Owen, W.L., Eds.; Springer: Boston, MA, USA, 2000; pp. 219–248. ISBN 978-1-4615-1401-5.
68. Horejsi, P.; Novikov, K.; Michal, Š. A Smart Factory in a Smart City: Virtual and Augmented Reality in a Smart Assembly Line. *IEEE Access* **2020**, *8*, 94330–94340. [CrossRef]
69. Di Pasquale, V.; De Simone, V.; Miranda, S.; Riemma, S. Smart Operators: How Augmented and Virtual Technologies Are Affecting the Worker's Performance in Manufacturing Contexts. *J. Ind. Eng. Manag.* **2022**, *15*, 233. [CrossRef]

machines

MDPI

Article

Correction of Shape Error at Cut-In and Cut-Out Points in Abrasive Waterjet Cutting of Carbon Fiber Reinforced Polymer (CFRP)

Ioan Alexandru Popan *, Cosmin Cosma, Alina Ioana Popan, Nicolae Panc, Daniel Filip and Nicolae Balc

Department of Manufacturing Engineering, Technical University of Cluj-Napoca, 400641 Cluj-Napoca, Romania; cosmin.cosma@tcm.utcluj.ro (C.C.); alina.luca@tcm.utcluj.ro (A.I.P.); nicolae.panc@tcm.utcluj.ro (N.P.); daniel.filip@mis.utcluj.ro (D.F.); nicolae.balc@tcm.utcluj.ro (N.B.)
* Correspondence: ioan.popan@tcm.utcluj.ro

Abstract: This paper presents a solution aimed at enhancing the accuracy of abrasive waterjet cutting (AWJC) for the processing of carbon-fiber-reinforced polymers (CFRP). Processing CFRP with high accuracy and good surface quality in a short processing time is a difficult task. One crucial problem is the occurrence of shape errors, overcuts, at the cut-in and cut-out point during the cutting process. Shape errors have the potential to create mechanical stress concentrators, which can result in structural failures and compromise the integrity and reliability of components. The primary objective of this study was to gain a comprehensive understanding of the formation mechanism underlying the shape error. The observed shape error is closely associated with both the lead-in/lead-out strategies employed and the process parameters selected. The experimental investigation focused on two commonly used strategies for CFRP cutting: lead-in/lead-out in arc and lead-in/lead-out in line. In order to minimize shape errors, this study proposed a correction method that offers a set of recommendations for selecting the appropriate lead-in/out strategy and a suitable combination of process parameters. Additionally, a mathematical model has been developed to determine the depth of the shape error. The conclusions drawn from this study have been successfully validated through industrial applications.

Keywords: abrasive waterjet cutting; lead-in/lead-out strategies; composite materials; CFRP

1. Introduction

In industries characterized by high standards and demands, such as the aerospace, automotive, and medicine industries, the utilization of advanced materials with exceptional properties is imperative [1,2]. Carbon-fiber-reinforced polymer/plastics (CFRP) are essential in high-end industries due to their exceptional properties. CFRP offers a high strength-to-weight ratio, making it strong and lightweight. Its excellent fatigue resistance ensures durability under repeated loading [1]. CFRP's stiffness and corrosion resistance make it stable and low-maintenance [2]. It maintains mechanical properties in extreme temperatures and provides design flexibility [3]. CFRP's biocompatibility is ideal for medical implants, and its impact resistance ensures safety in critical situations [4]. These properties have led to widespread adoption in various applications.

CFRPs have numerous aerospace applications due to their exceptional properties. CFRPs are extensively used in aircraft structures (fuselage sections, wings, tail sections), engine components (fan blades, spinner cones, nacelles, and exhaust component), interior components (seat frames, interior panels), helicopter rotor blades, spacecraft, satellites, and unmanned aerial vehicles (UAVs) [3,5,6].

The processing of advanced composite material with high accuracy and good surface quality in a short processing time is a difficult task [3,6]. The abrasive waterjet (AWJ) process is a suitable tool for cutting [7], turning [8], drilling [6], milling [9], cleaning [10], or surface preparation [11]. The AWJ processing principle involves the utilization of a high-pressure waterjet combined with abrasive particles to remove material, as is presented in Figure 1 [6,12]. The main AWJ parameters are waterjet pressure, abrasive mass flow, traverse speed or feed rate, standoff distance, and material thickness [12–14]. The kerf characteristics in abrasive waterjet (AWJ) cutting include several factors, including the kerf width, taper, top radius, and surface quality and integrity (as depicted in Figure 1b) [14–16].

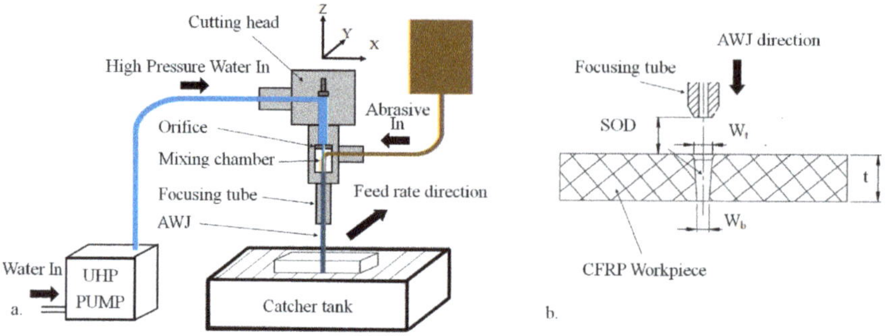

Figure 1. The AWJ process: (**a**) AWJ system and (**b**) kerf geometry.

Abrasive waterjet cutting (AWJC) uses a fluid tool suitable for machining all types of materials and it has several distinctive advantages such as low mechanical stress, complex shape-processing capabilities, an absence of heat-affected zones (HAZ) and thermal stress, and environmental-friendliness [9,12,13]. AWJC is also highly effective in cutting modern composite materials, whether they are produced using 3D printing technology from plastics or metals [17].

Defects such as delamination, fiber pull-out, abrasive embedment [6,15], dimensional deviation [7], or low surface quality [18] can occur during the processing of CFRP using AWJC technology. One of the most critical defects that can occur when machining CFRP with AWJ is delamination of the layers. The integrity and functionality of the part are compromised when this defect occurs [6,15,19,20]. According to Shanmugam et al. [20], delamination arises as a result of the shock wave generated when the waterjet impacts the material. Delamination typically occurs during the initial piercing phase rather than during the subsequent cutting process. To avoid delamination in CFRP during AWJC, several techniques can be employed: optimizing the cutting parameters, pre-drilling pilot holes, suitable nozzle configuration [6], or proper piercing methods [19]. The acoustical emission monitoring technique is employed to detect defects in CFRP, such as delamination, crack propagation, and fiber breakage [21,22].

The surface quality and integrity of CFRP after AWJC can significantly impact its overall performance and reliability [18,23]. Those directly affect its structural integrity. Any defects, delamination, or surface irregularities introduced during AWJC can weaken the material and compromise its load-bearing capacity. High-quality surfaces with minimal defects ensure a better structural stability and reliability [22]. CFRP materials are commonly used in applications where fatigue resistance is critical, such as in aerospace and automotive components. Poor surface quality, including micro-cracks and roughness, can act as stress concentrators, reducing the material's fatigue life and leading to premature failure [18,22]. In applications where aerodynamics or hydrodynamics are critical, such as in aerospace or marine structures, surface roughness and shape errors can lead to drag, turbulence, and performance inefficiencies. A smooth and clean-cut surface is desired to minimize post-processing requirements and maintain the structural integrity of the material. Achieving a good surface quality depends on factors like the cutting parameters, abrasive characteristics, cutting system, and material composition [18,22].

The characteristics of the kerf are investigated during the cutting process for CFRP [15]. The main geometrical kerf characteristics are width, taper, and top radius [14,15]. These geometrical characteristics directly impact the accuracy and overall quality of the cut parts. Proper selection and optimization of the process parameters is crucial for achieving high-quality cut surfaces and superior dimensional accuracy. The response surface methodology (RSM) is a suitable approach that can be employed for the optimization of AWJC [6,24].

Tolerances for CFRP parts can range from a few micrometers to a few millimeters, depending on the complexity of the part and the criticality of its function. The typical quality requirements in the case of the AWJC of composite materials used in aircraft structures are as follows: the surface roughness should not exceed Ra = 10 µm, the minimum accuracy required is about 0.25 mm, and the edge quality should have no adverse effects such as delamination [25]. For machining large aircraft structures such as wings or the fuselage, hybrid composite machining centers (combining AWJC and mechanical cutting processes) are utilized. These centers allow machining of parts up to 50 m in length with high accuracy [25].

An important aspect of the CFRP cutting process is the introduction of shape errors at the cut-in/cut-out point of the material. This shape error is in correlation with the lead-in/out strategy [7,26]. The lead-in/out strategy refers to a technique used to initiate and terminate the cutting process, as is shown in Figure 2. In this technique, specific strategies are employed to smoothly transition the AWJ into and out of the material being cut. The lead-in refers to the starting point (cut-in) of the waterjet on the material's surface, while the lead-out denotes the point where the AWJ exits the material.

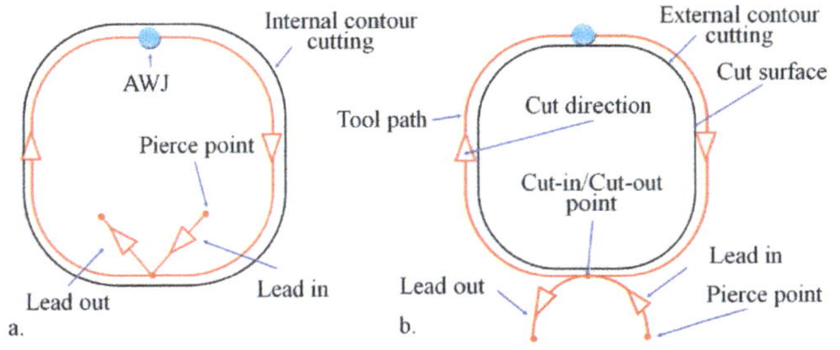

Figure 2. The principle of the lead-in/out strategy: (**a**) lead-in/out in a straight line and (**b**) lead-in/out in an arc.

The first recommendation is to work with outside corners, and it is usually preferable to extend the corner into a lead [27]. When processing an internal or external contour without outside corners (Figure 2), it is essential to employ the proper lead-in/lead-out strategies. There are various lead-in/lead-out strategies employed in AWJC:

- Lead-in/out in a straight line (Figure 2a): This strategy involves a linear or straight-line path for the waterjet during the cut-in and cut-out phases [26,27]. This strategy, also known as the "V" shape lead, is commonly preferred for general AWJC applications [27].
- Lead-in/out in an arc (Figure 2b): In this approach, the AWJ follows a curved or arc path during the cut-in and cut-out phases [27,28].

The choice of lead-in and lead-out strategies in AWJC can have a notable impact on the overall process time, especially in the context of industrial applications. A well-designed lead-in/out strategy can minimize the time spent on these movements, reducing the overall process time, and increasing productivity. Efficient lead-in/out strategies can optimize material utilization by minimizing waste.

When cutting CFRP with AWJ at the cut-in and cut-out points, there is a possibility of shape errors occurring, specifically overcutting [26]. These shape errors are also called witness marks [28]. AWJ is not a rigid tool as in the case of a milling cutter; it is a "soft knife" [7]. It moves to one side or the other of the tool path or stays behind the cutting head, generating a cutting lag [29]. The effects introduced by the AWJ are compensated for by the lead strategy and the acceleration and deceleration [26–28]. Ming Chen et al. conducted a study on shape errors occurring in the cut-in/cut-out region during the AWJC of Aluminum alloy 6061. It was concluded that the use of a very low feed rate during the lead-in/lead-out in line strategy contributed to the occurrence of shape errors. Although the errors were reduced, they were not completely eliminated [7,26].

Figure 3 illustrates an example of a shape error that occurred while cutting an internal contour in a 6 mm thick CFRP part. In Figure 3b, an overcut is observed at the cut-in and cut-out points. The overcut has a length of 2.2 mm and a depth of 1.3 mm.

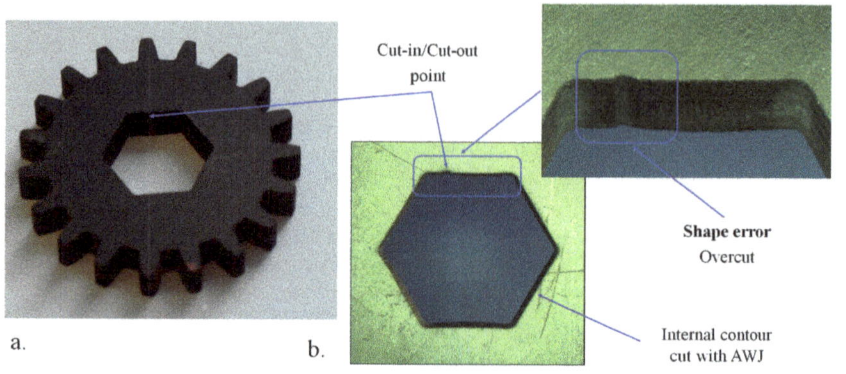

Figure 3. A shape error example appeared the cut-in and cut-out point. (**a**) A 6 mm CFRP part cut with AWJ. (**b**) The shape error (overcut) appeared during internal contour cutting.

Shape errors that result in mechanical stress concentrators can indeed lead to structural failures and compromise the integrity and reliability of components. These areas experience higher stress levels compared to the surrounding material, leading to potential weaknesses or failure points [23]. It is crucial to avoid the presence of mechanical stress concentrators to ensure the structural robustness of a design. By minimizing or eliminating stress concentrators, it can improve the overall strength, durability, and performance of the components, reducing the risk of failure and ensuring their long-term reliability [23,30]. Mechanical stress concentrators in aerospace CFRP components can have significant implications for their structural integrity and performance [30]. Franco et al. found that stress concentration around the drilled holes for fastening purposes in aluminum-composite hybrid joints used in aircraft structures could potentially lead to failure during flight due to fatigue loading [31]. Kant et al. conducted a comprehensive study of stresses in composite laminate structures, identifying significant stress concentrations at free edges and hole/cut-out/notch edges as potential factors driving failure, including the possibility of delamination [32]. Makki et al. conducted an evaluation of stress concentration factors for composite materials with varying hole diameters. Their experimental findings demonstrated an increase in gross stress concentration with an increase in hole diameter [33]. Ryo Naito et al. investigated the effects of stress concentration on CFRP behavior using a high-speed camera. Their study revealed that crack growth and shear failure occur from stress concentrator areas due to high stress concentration during the loading process just before final fracture [34].

When advanced composite materials such as CFRP are machined using AWJC, minimizing or eliminating the stress concentrators introduced by the cutting process becomes an important requirement.

The paper proposes a study aimed at solving the shape error problem introduced by AWJ during CFRP cutting. However, there is a research gap in understanding the specific influence of process parameters on shape errors at the cut-in and cut-out points in AWJC of CFRP. An experimental study was conducted to evaluate and understand the phenomenon. Two lead-in/lead-out strategies, lead-in/out in arc and lead-in/out in line, were investigated. The study aimed to determine the proper combination of strategy and process parameters. Based on the findings, a recommended strategy and set of process parameters were proposed.

2. Materials and Methods

2.1. Design of Experiments

In this experimental investigation, two different strategies, lead-in/out in arc and lead-in/out in line, were analyzed. To efficiently design the experiment and explore the effects of the variables, a factorial design was employed. Factorial designs allow for the simultaneous examination of multiple factors and their interactions [35]. Rammohan et al. employed a factorial design of experiments technique to predict AWJC parameters [36].

In Figure 4, the graphical representation of the analyzed strategies and the main geometrical parameters is presented.

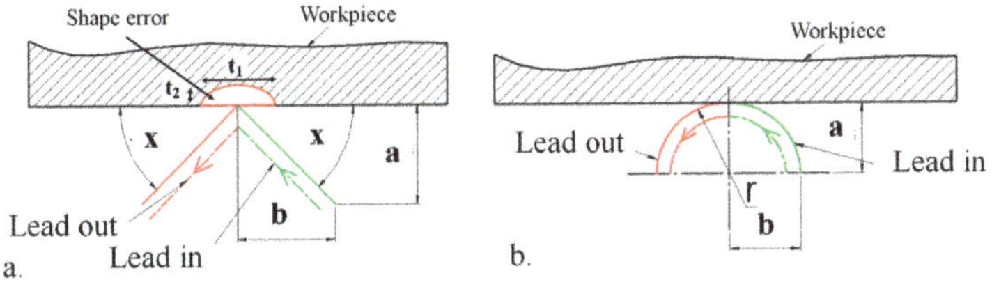

Figure 4. The graphical representation of the analyzed strategies: (**a**) lead-in/out in line and (**b**) lead -in/out in arc.

For the first strategy, which involves lead-in and lead-out in a straight line, the input variable selected was the lead angle (Figure 4a). The following values were chosen for the lead angle: 5, 35, 65, and 85°. Each lead line had a length of 3 mm. If the lead-in line is long enough, the influence of this parameter on the shape error is minimal [26].

In the case of lead-in and lead-out in an arc, the arc radius was varied with the following values: 1, 3, 5, and 7 mm (Figure 4b).

The nozzle traversing speed or feed rate, denoted as V, plays a key role in the AWJ process [15,22,36–38] and was varied throughout the experiment to assess its impact on the outcomes. In this study, the feed rate was calculated by using the Zeng mathematical model, Equation (1) [26,38].

$$V = \left(\frac{N_m P^{1.25} m_w^{0.68} m^{0.343}}{C_s Q H D^{0.618}} \right)^{1.15} \quad (1)$$

where V—feed rate (mm/min), N_m—machinability, P—waterjet pressure (MPa), m_w—water flow rate (L/min), m—abrasive flow rate (kg/s), C_s—measurement system constant, Q—surface quality, H—cutting thickness (mm), and D—internal focusing tube diameter (mm) [26,38].

This mathematical model is considered one of the most comprehensive and accurate models available. It has been widely adopted by commercial AWJ equipment manufacturers, such as Omax and Flow, who have integrated it into their software for calculating cutting parameters. The developed model includes a parameter called surface quality (Q), which is categorized into five classes (Q1–Q5) [39]. Figure 5 illustrates the quality characteristics of the cut surface.

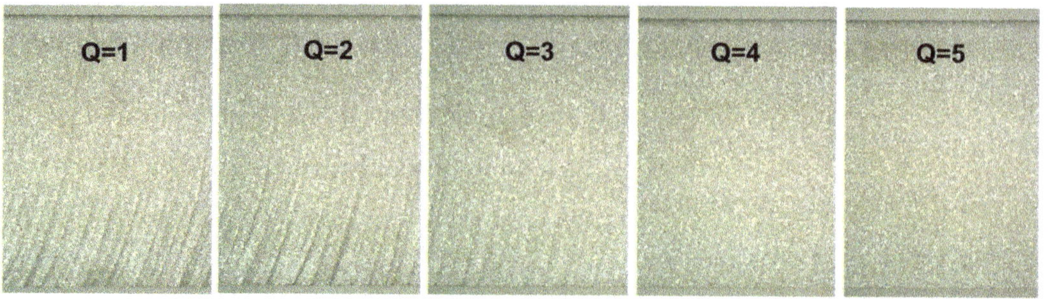

Figure 5. The quality characteristics of the cut surface.

The feed rate was calculated for three different cutting qualities: V = 4572 mm/min—Q1 separation quality, V = 4136 mm/min—Q3 medium quality, and V = 2299 mm/min—Q5 finishing quality. The constant process parameters were N_m = 500 (for CFRP), P = 345 MPa, m = 0.0075 kg/s, C_s = 788 (metric), H = 3 mm, and D = 0.76 mm.

During the experiment, a total of 72 trials were conducted. The experimental design followed a 4 × 3 full factorial design, resulting in 12 unique combinations of treatments [35]. Each treatment combination corresponds to a different combination of lead angle/arc and feed rate. To minimize processing and measurement errors, every experimental trial was repeated three times, ensuring reliable and accurate outcomes.

In Table 1, the input parameters of the AWJC process which were varied during the experiments are shown.

Table 1. The AWJC process input parameters, varied during the experiments.

Lead-In and Lead-Out Strategy:	Parameters	Value
Line	Lead angle (x) Feed rate (V)	5, 35, 65, 85° 2299, 4136, 4572 mm/min
Arc	Radius of the arc (r) Feed rate (V)	1, 3, 5, 7 mm 2299, 4136, 4572 mm/min

The output parameters analyzed in this study are the dimensions of shape error, specifically the depth (t_2) and width (t_1).

2.2. Experimental Setup

The experimental setup involved the use of an industrial AWJ machine, specifically the Omax 2626 model (Figure 6a). Omax machines are widely used in industries globally and share similarities with other brands like Flow, WJS, or Wardjet. The machine was equipped with a standard cutting head. The reason for selecting this machine is that the results obtained from this study can be easily implemented in industry, and the experiments can be replicated using commonly available equipment. A high-pressure pump was utilized to generate water pressure of up to a maximum of 400 MPa. To control the abrasive flow, an Alffi abrasive supply system was employed, allowing for variation in the abrasive mass flow within the range of 0 to 600 g/min.

A newly designed clamping device was developed to ensure the secure and rigid fixation of CFRP specimens during the cutting process. This clamping device is shown in Figure 6b. The CFRP specimens were machined using the AWJC process, resulting in final dimensions of 100 × 30 × 3 mm, as shown in Figure 6c.

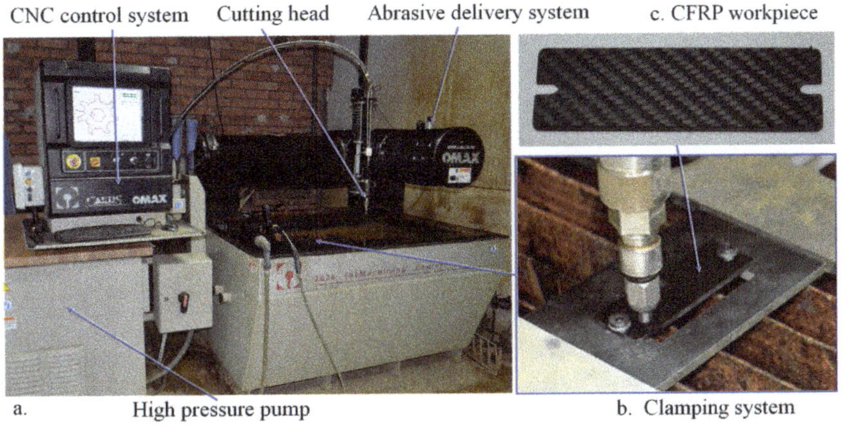

Figure 6. The experiment setup: (**a**) the Omax 2626 AWJ equipment, (**b**) the designed clamping system, and (**c**) the CFRP workpiece.

Table 2 presents the primary experimental parameters employed in this study. The cutting parameters were calculated for CFRP material cut with a thickness of 3 mm. To minimize the machining errors caused by the wear of the cutting system, a novel cutting system (focus tube and orifice) was installed.

Table 2. Experimental parameters.

Parameters	Value
Feed rate (V)	4572, 4136, 2299 mm/min
Waterjet pressure (P)	350 MPa
Standoff distance (SOD)	1 mm
Cutting depth (t)	3 mm
Abrasive mass flow (m_a)	0.45 kg/min
Focusing tube diameter	0.76 mm
Water nozzle diameter	0.35 mm

2.3. Materials

For this study, a 3 mm thick CFRP sheet manufactured by ECOTECH was used. The CFRP workpiece used in the study was produced using unidirectional prepreg carbon fiber (CE 8201-200-45) arranged at an angle of 0°/90°, combined with a 364 g/m^2 epoxy resin. The mechanical properties of the material were 230–294 GPa Young's modulus and 3530–5490 MPa tensile strength [40]. The production process involved hot pressing. The dimensions of the CFRP workpiece were 350 × 150 mm.

The abrasive grains used in the experimental trials were Barton garnet, a mineral material. Garnet is the most commonly used abrasive material in AWJC. The characteristics of the abrasive were as follows: a grain size of 260 μm, hardness of 7.5–8.0 Mohs, and specific density of 3.9–4.1 g/cm^3 [41].

2.4. Experimental Results

The shape error was measured optically using the PG 2000 Cutting Tool Inspection System, manufactured by Guhring. Figure 7a illustrates the setup of the microscope used for the measurement. This measuring microscope was well-suited for the size and shape of the samples. The measurements were taken from a perpendicular direction to the top surface of the specimen. With a 10× optical magnification, the measuring accuracy was approximately 2 μm. The shape error was evaluated by measuring two dimensions: t_2

depth of the shape error and t_1 length of the shape error (Figure 7b). A few machined samples are shown in Figure 7c,d.

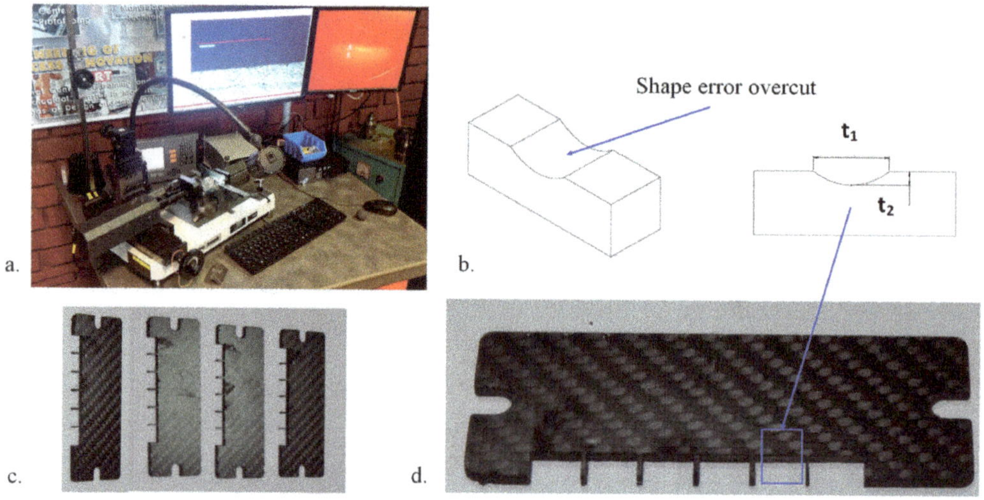

Figure 7. A process CFRP sample.

The experimental results obtained from the study are presented in Table 3, showcasing the outcomes and findings derived from the conducted experiments. The experimental trial was repeated three times, and the average of these values along with the standard deviation was calculated.

Table 3. The experimental results.

Feed Rate	**Shape Error**			**Lead-In/Lead-Out in Line**				**Lead-In/Lead-Out in Arc**			
				Lead Angle x (°)				**Radius of the Arc r (mm)**			
	Position Dimensions			**5**	**35**	**65**	**85**	**1**	**3**	**5**	**7**
Q1 separation quality **4572 mm/min**	Top	t_2 (mm)	Average	**0.049**	**0.084**	**0.090**	**0.103**	**0.057**	**0.056**	**0.046**	**0.047**
			St.dev.	0.0014	0.0109	0.009	0.0083	0.0018	0.0151	0.0023	0.0051
		t_1 (mm)	Average	**1.320**	**1.003**	**0.950**	**0.970**	**1.230**	**0.755**	**0.799**	**0.749**
			St.dev.	0.0065	0.0108	0.0076	0.0066	0.0175	0.0101	0.0095	0.013
	Bottom	t_2 (mm)	Average	**0.169**	**0.230**	**0.240**	**0.269**	**0.091**	**0.066**	**0.059**	**0.057**
			St.dev.	0.0077	0.0112	0.0056	0.0082	0.0103	0.0037	0.0122	0.0071
		t_1 (mm)	Average	**3.240**	**2.430**	**2.218**	**2.118**	**3.120**	**2.999**	**2.180**	**2.080**
			St.dev.	0.0066	0.0123	0.013	0.0032	0.0067	0.0035	0.0047	0.0052
Q3 medium quality **4136 mm/min**	Top	t_2 (mm)	Average	**0.036**	**0.064**	**0.066**	**0.076**	**0.050**	**0.042**	**0.043**	**0.040**
			St.dev.	0.0066	0.0029	0.0049	0.0103	0.0007	0.0022	0.0078	0.0031
		t_1 (mm)	Average	**1.191**	**0.821**	**0.900**	**0.850**	**0.860**	**0.699**	**0.500**	**0.560**
			St.dev.	0.0044	0.007	0.0131	0.0116	0.0043	0.0034	0.0142	0.0072
	Bottom	t_2 (mm)	Average	**0.140**	**0.191**	**0.203**	**0.215**	**0.089**	**0.060**	**0.053**	**0.055**
			St.dev.	0.0079	0.003	0.0083	0.0062	0.0026	0.0062	0.0051	0.0038
		t_1 (mm)	Average	**2.560**	**2.133**	**1.927**	**1.910**	**2.857**	**2.500**	**1.990**	**1.890**
			St.dev.	0.0156	0.0056	0.0026	0.0046	0.0091	0.0053	0.0064	0.0074

Table 3. *Cont.*

Feed Rate	Shape Error			Lead-In/Lead-Out in Line				Lead-In/Lead-Out in Arc			
				Lead Angle x (°)				Radius of the Arc r (mm)			
	Position	Dimensions		5	35	65	85	1	3	5	7
Q5 finishing quality **2299 mm/min**	Top	t_2 (mm)	Average	**0.030**	**0.045**	**0.060**	**0.058**	**0.042**	**0.037**	**0.036**	**0.038**
			St.dev.	0.0101	0.0093	0.0098	0.0051	0.0042	0.0047	0.0071	0.0049
		t_1 (mm)	Average	**1.120**	**0.751**	**0.670**	**0.690**	**0.710**	**0.458**	**0.430**	**0.400**
			St.dev.	0.0082	0.0045	0.0024	0.0091	0.0033	0.0037	0.007	0.0032
	Bottom	t_2 (mm)	Average	**0.110**	**0.165**	**0.189**	**0.198**	**0.086**	**0.059**	**0.051**	**0.049**
			St.dev.	0.0177	0.0104	0.0087	0.0038	0.0051	0.0082	0.0021	0.0032
		t_1 (mm)	Average	**2.423**	**1.830**	**1.736**	**1.650**	**2.810**	**1.985**	**1.836**	**1.736**
			St.dev.	0.0066	0.0155	0.0058	0.0039	0.0015	0.0061	0.0092	0.0083

3. Results and Discussions

The experimental data were analyzed to investigate the occurrence of shape errors at the cut-in and cut-out points during the cutting process. The depth of the shape error is a critical dimensional parameter that needs to be considered when evaluating the mechanical stress concentrator point.

3.1. Morphology of the Cut Surfaces

The cut surfaces of the CFRP specimens were examined using a microscope to study the behavior of the material during the AWJC process. This morphology analysis aimed to identify characteristics such as delamination, cracks, uncut fibers, fiber pull-out, or particle embedment [12,22,42,43]. The AWJ cut surfaces were examined and categorized into three zones: the initial damage zone (IDZ), the smooth cutting zone (SCZ), and the rough cutting zone (RCZ) [16].

In Figure 8a, microscopic images of a specimen obtained during this experiment are shown, highlighting the general shape error obtained ($t_{2\ bottom}$ = 0.203 mm in depth/ $t_{1\ bottom}$ = 1.92 mm). In this experimental run, the following parameters were used: lead-in/lead-out in line strategy (lead angle 65°), V = 4136 mm/min, P = 350 MPa, SOD = 1 mm, and m_a = 0.45 kg/min.

In the microscopic image from Figure 8b, a smooth surface quality is observed without any signs of matrix smearing, abrasive embedment, or a high percentage of pull-out fibers.

Surface roughness was measured using the Nano Focus 3D microscope µsurf expert, and a Gaussian filter with a 0.8 mm cutoff wavelength was applied. The dimensions of the scanned surface were 1.6 × 1.6 mm. The surface roughness was R_a =1.97 µm (S_a = 2.84 µm) on the IDZ, R_a =2.86 µm (S_a = 4.42 µm) on the SCZ, and for the RCZ, R_a =3.08 µm (S_a = 4.69 µm). Upon examining the topography and roughness profiles of the surfaces, it was noted that the valleys, peaks, and craters displayed an asymmetric and non-uniform pattern of variation. The surfaces display a wavy texture, characterized by the varying widths and heights of the peaks and valleys observed along their length. These peaks and valleys are aligned in the direction of the AWJ flow. Mardi et al. also observed similar surface topography characteristics on the surface cut with AWJ [43]. The maximum depth of the valleys is 6 µm, while the maximum height of the peaks is 5 µm. Additionally, a rounded edge (r_k = 0.12 mm) was observed on the IDZ.

The observations involved examining the cut surface at various magnifications to identify the CFRP damage (Figure 9). In 2 out of the 72 runs, delamination of the fiber layers was observed on the bottom part of the cut surface. In Figure 9b, a specimen is presented with delamination that occurred during a run utilizing the lead-in/lead-out in line strategy, with a lead angle of 5° and a feed rate of 4136 mm/min. In this case, the delamination measures 1.4 × 5 mm and is observed in the same area as the cut-in/cut-out point. Figure 9c showcases a different specimen with delamination. This delamination, measuring 0.8 × 3 mm, occurred in an experimental run employing the lead-in/lead-out in arc strategy with a lead radius of 3 mm and a cutting speed of 2299 mm/min. Notably, the cut-in/cut-out point of this run is in a different area compared to where the delamination occurred.

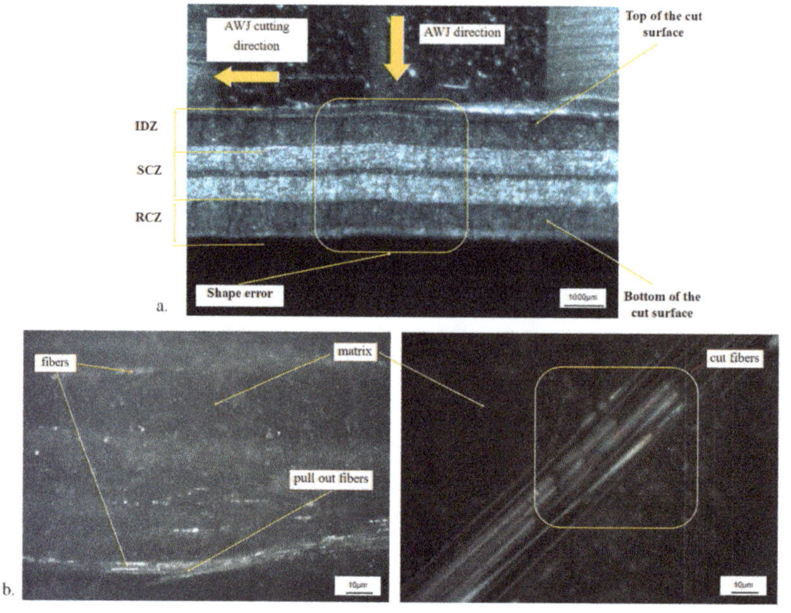

Figure 8. Microscopic image of the cut CFRP specimens.

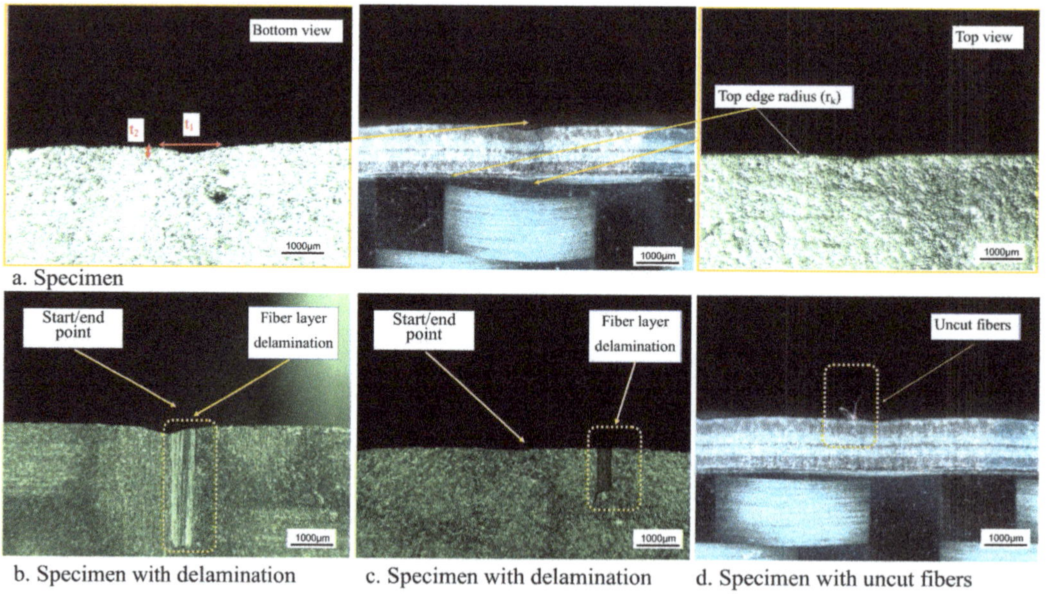

Figure 9. Morphological defects appear during experiments.

Uncut fibers appear in just one experimental run, in the rough cutting zone, shown in Figure 9d. It was obtained during a trial with the lead-in/lead-out in line strategy (lead angle 45°) and a feed rate of 2299 mm/min.

The defects, such as delamination of the fiber layers and uncut fibers, observed during the experimentation could potentially be attributed to defects in the CFRP material. However, after analysis, no specific correlation with process parameters or other factors was found. Further investigation is needed to better understand and address these issues.

This indicates that the cutting process was successful in preserving the structural integrity of the CFRP material without causing significant damage or defects.

3.2. Investigation on the Shape Error Generated by Using the Lead-In/Lead-Out in Line Strategy

Figure 10 illustrates the correlation between the shape error dimensions (depth t_2 and width t_1) and the process parameters (lead angle and feed rate) in the case of the lead-in/lead-out in line strategy. The shape error dimensions were evaluated for different values of lead angle (x: 5, 35, 65, 85°) and feed rate (V: 2299, 4136, 4572 mm/min).

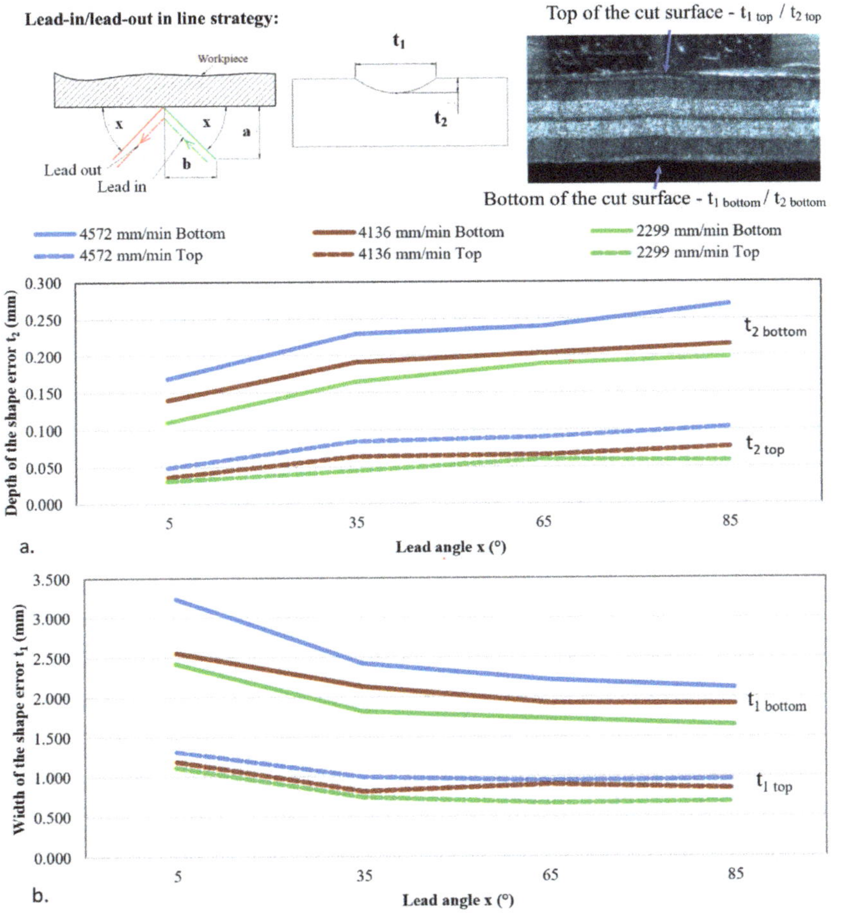

Figure 10. Influence of the lead angle on the shape error dimensions in the case of the lead-in/lead-out in line strategy: (**a**) the depth of the shape error t_2 and (**b**) the width of the shape error t_1.

In the case of the lead-in/lead-out in line strategy, the minimum dimensions of the shape error are, at the bottom, $t_{2\ bottom}$ = 0.11 mm in depth/$t_{1\ bottom}$ = 3.240 mm in width, and for the top of the cut surface, $t_{2\ top}$ = 0.03 mm in depth and $t_{1\ top}$ = 1.12 mm in width. This result was obtained during an experimental trial using the following parameters: lead angle x = 5°, V = 4572 mm/min, P = 350 MPa, SOD = 1 mm, and m_a = 0.45 kg/min. In the search to minimize the depth of the shape error, significant reductions were achieved, with a minimum depth of 0.11 mm being attained. However, complete elimination was not attained, as evidenced by the experimental findings.

As depicted in Figure 10a, it is evident that increasing the lead angle from 5° to 85° results in an increase in the depth of the shape error for both the top and bottom of the cut surface. Conversely, the width of the shape error decreases with an increase in the lead angle (Figure 10b).

When analyzing the effect of feed rate variation (2299–4572 mm/min), it was observed that decreasing the feed rate resulted in a decrease in the depth and width of the shape error (as shown in Figure 10).

The variation in shape error dimensions was analyzed in the case of the lead-in/lead-out in line strategy. The depth of the shape error was found to decrease with a decreasing feed rate and increase with an increasing lead angle. The width of the shape error decreases as the lead angle increases and increases with higher feed rates.

3.3. Investigation on the Shape Error Generated by Using the Lead-In/Lead-Out in Arc Strategy

Figure 11 presents the relationship between the shape error dimensions (depth t_2 and width t_1) and the process parameters (lead radius and feed rate) in the context of the lead-in/lead-out in arc strategy. The shape error dimensions were examined across various lead radius values (r: 1, 3, 5, 7 mm) and feed rate values (V: 2299, 4136, 4572 mm/min).

For lead-in/lead-out in arc strategies, the minimum dimensions of the shape error are, at the bottom, $t_{2\ bottom}$ = 0.049 mm in depth/$t_{1\ bottom}$ = 1.736 mm in width, and for the top of the cut surface, $t_{2\ top}$ = 0.038 mm in depth and $t_{1\ top}$ = 0.4 mm in width. It was obtained during an experimental trial using the following parameters: lead radius of 7 mm, V = 2299 mm/min, P = 350 MPa, SOD = 1 mm, and m_a = 0.45 kg/min. The selection criteria focused on minimizing the depth of the shape error. Although the depth of the shape error was reduced to below 0.05 mm, it was not completely eliminated.

Figure 11a demonstrates that increasing the lead arc radius from 1 to 7 mm results in a reduction in the depth of the shape error at both the bottom and top of the cut surface. Furthermore, Figure 11b indicates that as the lead arc radius increases, the width of the shape error decreases.

A decrease in the feed rate, ranging from 2299 to 4572 mm/min, resulted in a corresponding decrease in the depth and width of the shape error, as illustrated in Figure 10b. This consistent trend was observed for both the top and bottom of the cut surface.

In conclusion, for the lead-in/lead-out in arc strategy, to minimize the depth of the shape error, it is recommended to decrease the feed rate and increase the lead arc radius.

3.4. Discussions and Mathematical Modelling

To minimize or eliminate the stress concentrators that arise during the cutting process due to the lead-in/lead-out process, it is essential to gain a comprehensive understanding of the formation mechanism of the shape error.

The first observation indicates that the shape error (overcut) occurring at the cut-in and cut-out point AWJC of CFRP is significantly reduced but not completely eliminated. In lead-in/lead-out in arc strategies, the minimum dimensions of the shape error are observed at t_2 = 0.049 mm in depth and t_1 = 1.736 mm in width. On the other hand, in the case of the lead-in/lead-out in line strategy, the minimum dimensions are t_2 = 0.11 mm in depth and t_1 = 3.240 mm in width. When comparing these strategies based on the criterion of minimizing the depth of the shape error, it can be concluded that the lead-in/lead-out in arc strategies are significantly more accurate than the lead-in/lead-out in line strategy.

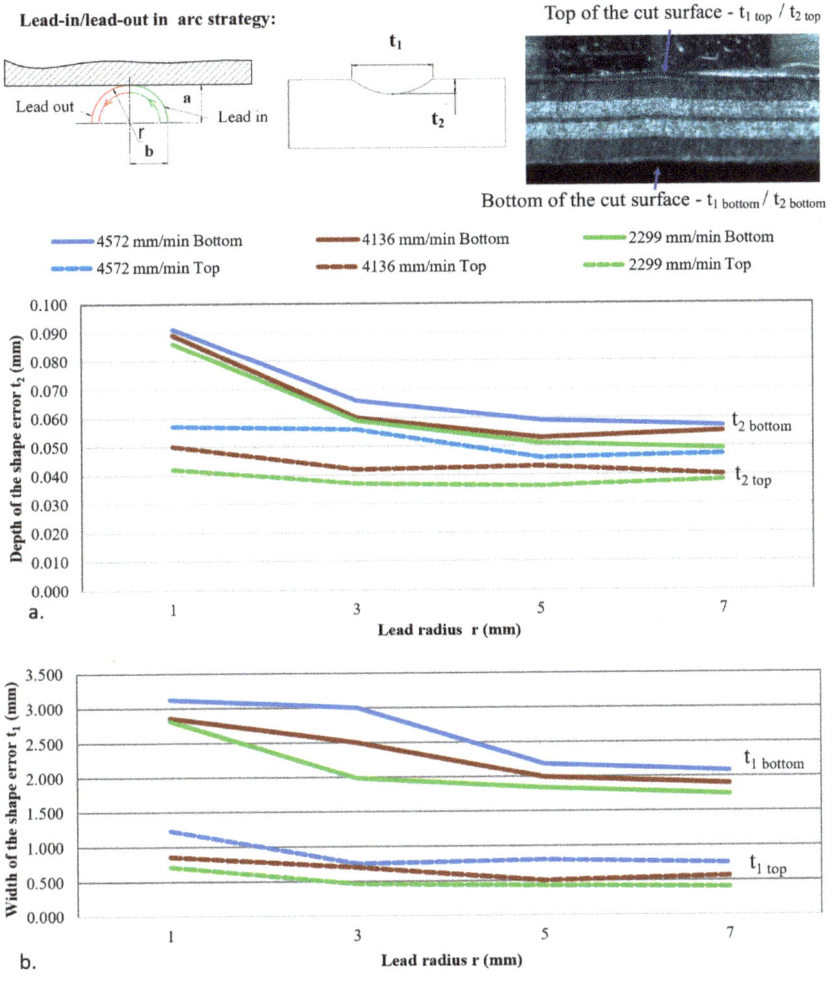

Figure 11. Influence of the lead radius on the shape error dimensions in the case of the lead-in/lead-out in arc strategy: (**a**) the depth of the shape error t_2 and (**b**) the width of the shape error t_1.

The shape error is in close correlation with the lead-in/lead-out geometry. CNC machines adapt the feed rate at direction changes by implementing advanced control algorithms and motion planning strategies. When a direction change occurs, the machine controller adjusts the feed rate to ensure a smooth and accurate motion [44].

The formation mechanism of the shape error is attributed to the deceleration of the machine during the lead-in and lead-out movements, which causes the AWJ to remove additional material from the workpiece. The decrease in the feed rate results in the generation of shape errors.

This theory is supported by experimental evidence. Increasing the lead arc radius from 1 to 7 mm generates a decrease in the depth of the shape error. Additionally, selecting a higher value for the lead arc radius (over 3 mm) ensures smoother jet movement without significant decelerations.

For lead-in/lead-out in line, as the lead angle increases from 5° to 85°, there is an increase in the reduction of the feed rate. To achieve a smooth and constant movement, it is recommended to decrease the lead angle, which also leads to a decrease in the depth of the shape error.

In conclusion, to minimize the depth of the shape error, it is recommended to select the lead-in/lead-out in arc strategy with a high value of lead arc radius (over 3 mm). This approach will result in smoother jet movement without significant decelerations.

A decrease in the feed rate, ranging from 4572 to 2299 mm/min, led to a reduction in the depth and width of the shape error. This trend was observed for both lead-in/lead-out strategies. By reducing the feed rate, the shape error is decreased due to a lower percentage of deceleration during the lead-in/lead-out process. This reduction in the feed rate not only leads to a smoother and more accurate motion but also results in a reduced variation of the feed rate along the cutting tool path.

It was seen that reducing the feed rate resulted in improved surface roughness and accuracy. Additionally, it was found that the kerf taper decreased as the feed rate decreased. The observation made by Popan et al. aligns with the findings of this study [22].

Multiple models were trained using the experimental data to predict the depth of the shape error in AWJC. The experimental results obtained in the AWJC process are suitable for statistical analysis using ANOVA (analysis of variance) [6,45]. It is a powerful statistical tool that helps us understand the relationship between variations in the dependent variable (the depth of the shape error) and different independent variables (feed rate and lead radius).

In this study, several models were tested to predict the shape error depth during the AWJC process. These models included cubic, two-factor interaction, quadratic, and linear models. The fit test was conducted to evaluate how well each model fits the experimental data. Among all the models tested, the linear model showed a statistically significant relationship with the data ($p < 0.05$, $p = 0.0004$). This means that the linear model provides a strong and meaningful representation of the relationship between the process parameters and the shape error depth. The significance level of $p = 0.0004$ indicates that the relationship is highly unlikely to occur by chance, further confirming the validity of the linear model in explaining the observed data.

The statistical analysis, presented in Table 2, demonstrated that the linear model provides a good fit to the data, with an adjusted R-squared value of 0.9081. This high R-squared value indicates that approximately 91% of the variability in the dependent variable can be explained by the linear model, making it a reliable predictor for the given data.

In this model, the lead radius is a significant predictor ($p = 0.0004$) of the depth of the shape error (Table 4). The regression model equation derived from the analysis is as follows:

$$t_2 = 0.0748 - 0.0056 \times r + 2.95 \times 10^{-6} \times V \quad (2)$$

where t_2—depth of shape error (mm), r—lead radius (mm), and V—feed rate (mm/min).

Table 4. Linear regression model statistics.

Multiple R		R Square	Adjusted R Square	Standard Error	
0.9529		0.9081	0.7868	0.0212	
ANOVA	**df**	**SS**	**Mean Square**	**F**	**Significance F**
Regression	2	0.04022	0.02012	44.5108	0.00046
Residual	9	0.00406	0.00045		
Total	11	0.04432			
		Coefficients	**Standard Error**	**t Stat**	***p*-Value**
Intercept		0.0748	0.00952	7.84962	5.00465×10^{-5}
Lead radius (r)		−0.0056	0.00121	−4.60739	0.00173
Feed rate (V)		2.95×10^{-6}	2.71438×10^{-6}	1.08727	0.30858

The model was completed by incorporating the feed rate calculation model developed by Zeng (Equation (1)). By integrating Zeng's model into the framework, the depth of the shape error can be calculated, considering important parameters such as water pressure, material thickness, abrasive mass flow, and cutting quality. This results in the formulation of a new model, represented by Equation (3), which enables the estimation of the depth of the shape error based on the process parameters.

$$t_2 = 0.0748 - 0.0056 \times r + 2.95 \times 10^{-6} \times \left(\frac{N_m P^{1.25} m_w^{0.68} ma^{0.343}}{C_s Q H D^{0.618}} \right)^{1.15} \quad (3)$$

where t_2—depth of shape error (mm), r—lead radius (mm), N_m—machinability (500 for CFRP), P—waterjet pressure (MPa), m_w is the water flowrate (L/min), m_a—abrasive flow rate (kg/s), Cs—measurement system constant (for metric is 788), Q—surface quality (Q1 separation, Q3 medium, Q5 finishing), H—cutting depth (mm), and D—internal diameter of the focusing tube (mm) [26,38].

The model was experimentally validated through thirty trials using composite material (CFRP) with thicknesses of 2, 4, and 6 mm. The results demonstrated a satisfactory level of agreement between the experimental values and the calculated values, with a maximum difference of 14%.

3.5. Shape Error Correction Method

The shape error correction method involves a set of recommendations for selecting the appropriate lead-in/out strategy and a suitable combination of process parameters. To minimize the depth of the shape error at the cut-in and cut-out points during the abrasive waterjet cutting of composite materials like CFRP, the following recommendations are advised:

- The lead-in/lead-out in arc strategy should be chosen, as it facilitates smooth AWJ movement and helps to maintain a constant feed rate. It is recommended to increase the lead arc radius to a value greater than 3 mm.
- To minimize the depth of the shape error, it is essential to find the optimal combination of process parameters. One key parameter is the feed rate or traverse speed. It is recommended to decrease the feed rate, as this helps decrease the depth of the shape error. The feed rate can be determined using the Zeng model, which takes into account the key parameters of the AWJC process (Equation (1)). In this application, the finishing cutting quality Q5 is recommended.
- The model that was formulated (Equation (3)) can be employed to estimate the depth of the shape error.

3.6. Method Validation

The proposed methodology was implemented in an industrial company for the cutting of 120 parts made of CFRP with a thickness of 5 mm. The external contour (225 × 72 mm) and five holes with a diameter of 20 ± 0.05 mm were machined, as shown in Figure 12.

The cutting tests were conducted using an AWJC CNC machine, CMS Tecnocut 1740, which was equipped with a 60 HP hydraulic pump capable of generating pressures up to 400 MPa. The machine's movements were controlled by a TechnoCAM CNC controller, which operates in three axes.

Based on the shape error correction method, the lead-in/lead-out strategy was selected. Lead-in/lead-out in arc strategies were implemented with a 5 mm arc radius value. To minimize the depth of the shape error, it is recommended to increase the lead arc radius to a value greater than 3 mm. However, this value should not be increased too much as it could lead to a longer tool path and increased machining time. An optimal value must be selected. The feed rate value, V = 2050 mm/min, was calculated using Equation (1) to achieve a finishing cutting quality (Q5). The cutting parameters utilized were as follows: P = 340 MPa, V = 2050 mm/min, SOD = 1 mm, and m_a = 0.45 kg/min. The holes were initially pierced using low pressure (P = 100 MPa) with the abrasive waterjet (AWJ) to prevent material delamination.

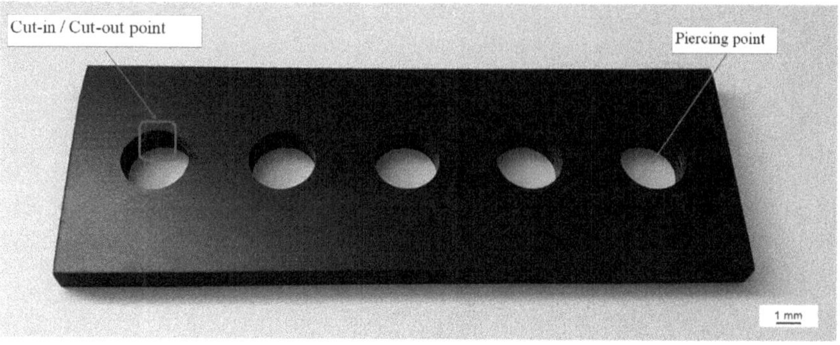

Figure 12. The CFRP industrial part processed with the proposed method.

The part was successfully cut without any delamination issues, meeting the required quality standards. The dimensions of the shape error were measured using an optical microscope, following the same procedure as the previous experiment. The measurements revealed a depth of 0.04 mm and a width of 1.9 mm at the bottom of the cut surface. For the top of the cut surface, the depth was measured to be 0.055 mm (0.052 mm calculated value with the developed model) and the width was 0.6 mm. The depth of the shape error was reduced to 0.055 mm from its initial value of 0.3 mm before the implementation of the proposed methodology. The implementation of the new method resulted in a significant reduction in the shape error by 81%. The machining time for the part using the newly implemented method was 1.8 min, while the previous method resulted in a machining time of 1.85 min. A reduction in time was obtained, 0.05 min, suggesting that the implementation of the new method has influenced the machining time. The machining time is strongly influenced by the length of the lead-in/out tool path, and in this validation test, those lengths were similar.

4. Conclusions

This paper presents an investigation focused on reducing the shape errors that occur at the cut-in and cut-out points during abrasive waterjet cutting of CFRP. Shape errors that create areas of concentrated mechanical stress have the potential to cause structural failures and compromise the integrity and reliability of CFRP components. Therefore, it is crucial to minimize or eliminate stress concentrators that are introduced during the cutting process. The research yielded the following conclusions:

- The shape error observed during the cutting process is caused by the decrease in the feed rate at the cut-in and cut-out points. This is attributed to the machine deceleration during lead-in and lead-out movements, resulting in the removal of extra material from the workpiece.
- Although the shape error could be significantly reduced, it was not completely eliminated. In the case of lead-in/lead-out in arc strategy, the minimum dimensions of the shape error were found to be 0.049 mm in depth and 1.736 mm in width. Conversely, for lead-in/lead-out in line strategy, the minimum dimensions were 0.11 mm in depth and 3.240 mm in width. Based on the criterion of minimizing the depth of the shape error, the lead-in/lead-out in arc strategy was found to be more suitable for CFRP cutting.
- A shape error correction method was proposed, providing a set of recommendations for selecting the appropriate lead-in/out strategy and a suitable combination of process parameters. This approach offers several benefits, including minimizing the shape

error (up to 80%), improving overall cutting accuracy (up to ±0.05 mm), and reducing the machining time.
- The proposed method has been successfully validated and implemented in industrial applications, demonstrating its effectiveness in improving the accuracy of CFRP cutting and reducing the machining time. This research contributes to the advancement of AWJC technology and provides valuable insights for practical applications in the manufacturing industry.
- A mathematical model was formulated to estimate the depth of the shape error, considering the main process parameters of AWJC. This model provides a predictive tool with over 85% accuracy, enabling the estimation of the shape error depth and optimization of the cutting process.
- Future investigations could explore new composite materials (reinforced with fiberglass, kevlar, wood, metal), optimize the jet trajectory by involving multi-axis AWJC machines, or design new geometries for lead-in/lead-out tool paths, such as spline curves.

Author Contributions: Conceptualization, methodology, investigation, I.A.P., A.I.P., and C.C.; validation, I.A.P. and A.I.P.; supervision, N.B. and D.F.; writing—original draft, I.A.P., C.C., and N.P.; research funding, N.B. and D.F. All authors have read and agreed to the published version of the manuscript.

Funding: This research was supported by HORIZON 2020—DiCoMI Project, "Directional Composites through Manufacturing Innovation", GA Nr. 778068.

Data Availability Statement: Not applicable.

Conflicts of Interest: The authors declare no conflict of interest.

References

1. Ding, J.; Cheng, L.; Chen, X.; Chen, C.; Liu, K. A Review on Ultra-High Cycle Fatigue of CFRP. *Compos. Struct.* **2021**, *256*, 113058. [CrossRef]
2. Al-Lami, A.; Hilmer, P.; Sinapius, M. Eco-Efficiency Assessment of Manufacturing Carbon Fiber Reinforced Polymers (CFRP) in Aerospace Industry. *Aerosp. Sci. Technol.* **2018**, *79*, 669–678. [CrossRef]
3. Poór, D.I.; Geier, N.; Pereszlai, C.; Xu, J. A Critical Review of the Drilling of CFRP Composites: Burr Formation, Characterisation and Challenges. *Compos. Part B Eng.* **2021**, *223*, 109155. [CrossRef]
4. Chua, C.Y.X.; Liu, H.-C.; Di Trani, N.; Susnjar, A.; Ho, J.; Scorrano, G.; Rhudy, J.; Sizovs, A.; Lolli, G.; Hernandez, N.; et al. Carbon Fiber Reinforced Polymers for Implantable Medical Devices. *Biomaterials* **2021**, *271*, 120719. [CrossRef]
5. Guo, F.; Xiao, Q.; Xiao, S.; Wang, Z. Assembly Technology for Aeronautical CFRP Structures under the Collaborative Constrains of Geometric Shape, Physical Performance and Service Stability. *Compos. Struct.* **2023**, *318*, 117071. [CrossRef]
6. Popan, I.A.; Balc, N.; Popan, A.I. Avoiding Carbon Fibre Reinforced Polymer Delamination during Abrasive Water Jet Piercing: A New Piercing Method. *Int. J. Adv. Manuf. Technol.* **2022**, *119*, 1139–1152. [CrossRef]
7. Chen, M.; Zhang, S.; Zeng, J.; Chen, B.; Xue, J.; Ji, L. Correcting Shape Error on External Corners Caused by the Cut-in/Cut-out Process in Abrasive Water Jet Cutting. *Int. J. Adv. Manuf. Technol.* **2019**, *103*, 849–859. [CrossRef]
8. Hutyrová, Z.; Ščučka, J.; Hloch, S.; Hlaváček, P.; Zeleňák, M. Turning of Wood Plastic Composites by Water Jet and Abrasive Water Jet. *Int. J. Adv. Manuf. Technol.* **2015**, *84*, 1615–1623. [CrossRef]
9. Dekster, L.; Karkalos, N.E.; Karmiris-Obratański, P.; Markopoulos, A.P. Evaluation of the Machinability of Ti-6Al-4V Titanium Alloy by AWJM Using a Multipass Strategy. *Appl. Sci.* **2023**, *13*, 3774. [CrossRef]
10. Hardy, M.C.; Herbert, C.R.J.; Kwong, J.; Li, W.; Axinte, D.A.; Sharman, A.R.C.; Encinas-Oropesa, A.; Withers, P.J. Characterising the Integrity of Machined Surfaces in a Powder Nickel Alloy Used in Aircraft Engines. *Procedia CIRP* **2014**, *13*, 411–416. [CrossRef]
11. Nag, A.; Srivastava, M.; Petrů, J.; Váňová, P.; Srivastava, A.K.; Hloch, S. Comparison of Continuous and Pulsating Water Jet during Piercing of Ductile Material. *Materials* **2023**, *16*, 3558. [CrossRef] [PubMed]
12. Srinivasu, D.S.; Axinte, D.A. Surface Integrity Analysis of Plain Waterjet Milled Advanced Engineering Composite Materials. *Procedia CIRP* **2014**, *13*, 371–376. [CrossRef]
13. Zou, X.; Fu, L.; Wu, L.; Zuo, W. Research on Multiphase Flow and Nozzle Wear in a High-Pressure Abrasive Water Jet Cutting Head. *Machines* **2023**, *11*, 614. [CrossRef]
14. Basarman, A.-P.; Lobonțiu, M. Kerf Variation Analysing for Abrasive Water Jet Cutting of a Steel Square Part. *MATEC Web Conf.* **2017**, *112*, 03002. [CrossRef]

15. Dhanawade, A.; Kumar, S. Experimental Study of Delamination and Kerf Geometry of Carbon Epoxy Composite Machined by Abrasive Water Jet. *J. Compos. Mater.* **2017**, *51*, 3373–3390. [CrossRef]
16. Popan, I.A.; Contiu, G.; Campbell, I. Investigation on Standoff Distance Influence on Kerf Characteristics in Abrasive Water Jet Cutting of Composite Materials. *MATEC Web Conf.* **2017**, *137*, 01009. [CrossRef]
17. Szatkiewicz, T.; Perec, A.; Radomska-Zalas, A.; Banaszek, K.; Balasz, B. Preliminary Studies into Cutting of a Novel Two Component 3D-Printed Stainless Steel–Polymer Composite Material by Abrasive Water Jet. *Materials* **2023**, *16*, 1170. [CrossRef]
18. Bañon, F.; Sambruno, A.; Batista, M.; Simonet, B.; Salguero, J. Study of the Surface Quality of Carbon Fiber–Reinforced Thermoplastic Matrix Composite (CFRTP) Machined by Abrasive Water Jet (AWJM). *Int. J. Adv. Manuf. Technol.* **2020**, *107*, 3299–3313. [CrossRef]
19. Kim, G.; Denos, B.R.; Sterkenburg, R. Influence of Different Piercing Methods of Abrasive Waterjet on Delamination of Fiber Reinforced Composite Laminate. *Compos. Struct.* **2020**, *240*, 112065. [CrossRef]
20. Shanmugam, D.K.; Nguyen, T.; Wang, J. A Study of Delamination on Graphite/Epoxy Composites in Abrasive Waterjet Machining. *Compos. Part A Appl. Sci. Manuf.* **2008**, *39*, 923–929. [CrossRef]
21. Lissek, F.; Haeger, A.; Knoblauch, V.; Hloch, S.; Pude, F.; Kaufeld, M. Acoustic Emission for Interlaminar Toughness Testing of CFRP: Evaluation of the Crack Growth Due to Burst Analysis. *Compos. Part B Eng.* **2018**, *136*, 55–62. [CrossRef]
22. Popan, I.A.; Bocanet, V.; Balc, N.; Popan, A.I. Investigation on Feed Rate Influence on Surface Quality in Abrasive Water Jet Cutting of Composite Materials, Monitoring Acoustic Emissions. In *Advances in Manufacturing Engineering and Materials*; Lecture Notes in Mechanical, Engineering; Hloch, S., Klichová, D., Krolczyk, G.M., Chattopadhyaya, S., Ruppenthalová, L., Eds.; Springer International Publishing: Cham, Switzerland, 2019; pp. 105–113. ISBN 978-3-319-99352-2.
23. Yamamoto, G.; Onodera, M.; Koizumi, K.; Watanabe, J.; Okuda, H.; Tanaka, F.; Okabe, T. Considering the Stress Concentration of Fiber Surfaces in the Prediction of the Tensile Strength of Unidirectional Carbon Fiber-Reinforced Plastic Composites. *Compos. Part A Appl. Sci. Manuf.* **2019**, *121*, 499–509. [CrossRef]
24. Perec, A.; Fajdek-Bieda, A.; Pude, F.; Radomska-Zalas, A. Process optimization by applying the response surface methodology (RSM) to the abrasive suspension water jet cutting of phenolic composites. *Facta Univ.* **2022**, *20*, 1–16. [CrossRef]
25. Mohamed, H. Trimming of CFRP Aircraft Components. In Proceedings of the WJTA-IMCA Conference and Expo, Houston, TX, USA, 9–11 September 2013.
26. Chen, M.; Zhang, S.; Zeng, J.; Chen, B. Correcting Shape Error Located in Cut-in/Cut-out Region in Abrasive Water Jet Cutting Process. *Int. J. Adv. Manuf. Technol.* **2019**, *102*, 1165–1178. [CrossRef]
27. Omax "Precision Practices: Lead-Ins and Lead-Outs". 2023. Available online: https://www.Omax.Com/En/Us/Media-Center/Tips/Precision-Practices-Lead-Ins-and-Lead-Outs (accessed on 1 June 2023).
28. WARDJet Reducing Lead-in/Lead-out Witness Marks. 2023. Available online: https://my.wardjet.com/Waterjet/University/Precision-Quality (accessed on 1 June 2023).
29. Hlaváč, L.M. Revised Model of Abrasive Water Jet Cutting for Industrial Use. *Materials* **2021**, *14*, 4032. [CrossRef] [PubMed]
30. Kumar, S.A.; Rajesh, R.; Pugazhendhi, S. A Review of Stress Concentration Studies on Fibre Composite Panels with Holes/Cutouts. *Proc. Inst. Mech. Eng. Part L J. Mater. Des. Appl.* **2020**, *234*, 1461–1472. [CrossRef]
31. Di Franco, G.; Zuccarello, B. Analysis and Optimization of Hybrid Double Lap Aluminum-GFRP Joints. *Compos. Struct.* **2014**, *116*, 682–693. [CrossRef]
32. Kant, T.; Swaminathan, K. Estimation of Transverse/Interlaminar Stresses in Laminated Composites—A Selective Review and Survey of Current Developments. *Compos. Struct.* **2000**, *49*, 65–75. [CrossRef]
33. Mohamed Makki, M.; Chokri, B. Experimental, Analytical, and Finite Element Study of Stress Concentration Factors for Composite Materials. *J. Compos. Mater.* **2017**, *51*, 1583–1594. [CrossRef]
34. Ryo, N.; Mitsuhiro, O.; Daisuke, F. Effects of Stress Concentration on the Mechanical Properties of Carbon Fiber Reinforced Plastic. *Int. J. Smart Mater. Mechatron.* **2015**, *2*, 136–139.
35. Montgomery, D.C. *Design and Analysis of Experiments*, 10th ed.; Wiley: Hoboken, NJ, USA, 2020; ISBN 978-1-119-49247-4.
36. Rammohan, S.; Kumaran, S.T.; Uthayakumar, M.; Korniejenko, K.; Nykiel, M.; Velayutham, A. Prediction of Abrasive Waterjet Machining Parameters of Military-Grade Armor Steel by Semi-Empirical and Regression Models. *Materials* **2022**, *15*, 4368. [CrossRef] [PubMed]
37. Jankovic, P.; Igic, T.; Nikodijevic, D. Process Parameters Effect on Material Removal Mechanism and Cut Quality of Abrasive Water Jet Machining. *Theor. Appl. Mech.* **2013**, *40*, 277–291. [CrossRef]
38. Zeng, J.; Kim, T.J. Development of an Abrasive Waterjet Kerf Cutting Model for Brittle Materials. In *Jet Cutting Technology*; Springer: Berlin/Heidelberg, Germany, 1992; pp. 483–501.
39. Valleyknife. Available online: http://www.valleyknife.com/waterjet_services.html (accessed on 1 June 2023).
40. ECOTECH—CFRP Manufacturer. Available online: https://www.r-g.de/en/index.html (accessed on 26 May 2023).
41. Nag, A.; Ščučka, J.; Hlavacek, P.; Klichová, D.; Srivastava, A.K.; Hloch, S.; Dixit, A.R.; Foldyna, J.; Zelenak, M. Hybrid Aluminium Matrix Composite AWJ Turning Using Olivine and Barton Garnet. *Int. J. Adv. Manuf. Technol.* **2018**, *94*, 2293–2300. [CrossRef]
42. Thongkaew, K.; Wang, J.; Li, W. An Investigation of the Hole Machining Processes on Woven Carbon-Fiber Reinforced Polymers (CFRPs) Using Abrasive Waterjets. *Mach. Sci. Technol.* **2019**, *23*, 19–38. [CrossRef]
43. Mardi, K.B.; Dixit, A.R.; Pramanik, A.; Hvizdos, P.; Mallick, A.; Nag, A.; Hloch, S. Surface Topography Analysis of Mg-Based Composites with Different Nanoparticle Contents Disintegrated Using Abrasive Water Jet. *Materials* **2021**, *14*, 5471. [CrossRef]

44. Yang, M.; Yang, J.; Zhu, L.; Yu, X. A Novel Curvature Circle Iterative Algorithm for Contour Error Control of Multi-Axis CNC Machine Tools. *Precis. Eng.* **2020**, *65*, 23–31. [CrossRef]
45. Mohankumar, V.; Kanthababu, M.; Velayudham, A. Abrasive Waterjet Cutting of Boron Carbide Particles Reinforced Al 6063 MMCs—A Semi Empirical Modeling Approach in the Prediction of Kerf Angle. *Measurement* **2021**, *181*, 109492. [CrossRef]

machines

MDPI

Article

Simulation Design and Measurement of Welding Robot Repeatability Utilizing the Contact Measurement Method

Martin Pollák * and Karol Goryl

Faculty of Manufacturing Technologies, Technical University of Kosice, Bayerova 1, 080 01 Presov, Slovakia; karol.goryl@tuke.sk
* Correspondence: martin.pollak@tuke.sk; Tel.: +421-55-602-6460

Abstract: ISO 9283 is a significant guiding standard for assessing the performance characteristics of robots. The main objective of the present paper was to verify the repeatability of the Panasonic TM-2000 welding robot at a manufacturing company. The paper describes the workflow of the robot control program in the simulation software RoboDK, which created a complete welding station. The measurement, as well as the simulation, were possible thanks to designing the measuring device, the imaginary ISO cube, the measuring plane, the measuring points, and last but not least, the cycle that determined in which order the points were measured. The analytical part of the paper resulted in a direct measurement of the position repeatability of the welding robot. A total of five points were measured for the X-axis and five points for the Y-axis. Each point was recorded 30 times, with measurements taken in the positive direction of motion. The results were compared with the value given by the manufacturer, and the measured deviations were presented graphically.

Keywords: ISO 9283 standard; unidirectional repeatability; positioning accuracy; RoboDK simulation software; Panasonic TM-2000 welding robotic arm

1. Introduction

Robotics can be applied to not only in large engineering companies, but thanks to its availability, it can also reach the medium and smaller manufacturing companies. The advantage of using robots is their complexity and adaptability within the manufacturing process. After a certain period, a robot that performs work directly in the manufacturing process may lose its declared accuracy and repeatability for various reasons, such as an arm overload or irregular calibration [1]. The main performance characteristics of the robots to be monitored are the repeatability, payload, reach, and positioning accuracy (Figure 1) [2,3].

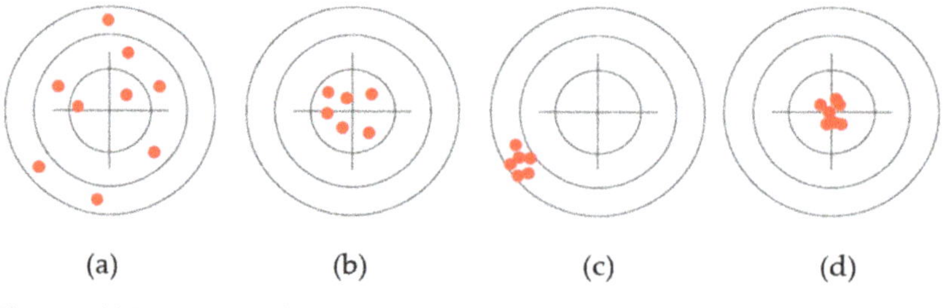

Figure 1. (**a**) Low repeatability and accuracy. (**b**) Low repeatability and high accuracy. (**c**) High repeatability and low accuracy. (**d**) High repeatability and accuracy.

Payload is defined as the weight that the robot can carry without causing variations in other specifications, such as accuracy, during the execution of the task. The maximum payload of the robotic arm can be greater than the specified payload. At a load greater than the useful load, the robot will likely be less accurate, experience a problem with the trajectory of the motion, or have large deviations [4]. An example is the specification of a robot where the maximum payload specified by the manufacturer is 6 kg, which means that the robot has been tested at 6 kg, the weight at which it has achieved the best results. This weight should not be exceeded to avoid positioning errors [5].

Reach can be defined as the maximum distance the robot can reach within its working area [6]. The cubes' function, the robot arm's length, and other robot configurations condition the robot's reach. The parameters provided by the manufacturer only give the maximum range, which is a rough estimate of the workspace. In practice, it is common for a robot to have a greater reach in one direction than in the other. Therefore, the robot's reach is very important in terms of robot selection [7,8].

Accuracy is generally defined as the error between the programmed, i.e., desired, position and the actual position of the robot endpoint. Accuracy is not always paramount when using a robot, meaning that if the robot is programmed manually, where the worker teaches the robot all the points to reach, accuracy does not play a major role. However, the positioning accuracy of the robot is particularly important when it is used in a market where it has to be programmed according to a CAD model, as the endpoint has to be reached in precisely defined positions [9].

The repeatability of a robot expresses how accurately the robot can reach the same position if the motion is repeated several times. The repeatability of a position is affected by the robot's precision. For example, when testing a robot where a desired position is selected and repeated 30 times, the robot will not always reach the same position. However, deviations will occur, and the robot will move near the desired point. The radius that is formed by the repeated movements is termed the repeater. A certain number of repetitive cycles usually form this robot characteristic. The more repetitions that are performed, the worse the results obtained. Robot manufacturers specify the tests they have performed with a certain number of cycles, payload, and arm orientation [10].

2. State of the Art

The problems in evaluating the performance characteristics of robots, such as unidirectional precision and the robot's position, is currently being addressed by many research teams and manufacturing departments worldwide. The measurement itself can take place directly in the manufacturing plants, as well as in the laboratory columns.

In an experiment conducted by Sepehr Gharaaty et al., the positioning accuracy of two industrial robots, specifically the FANUC LR Mate 200iC and the FANUC M20iA, was investigated. The authors solved the problem by designing a dynamic position correction scheme. This dynamic position correction scheme works by obtaining a feedback loop through which the change in positioning is through the projected robotic arm. The authors used the C-Track 780 optical coordinate system from Creaform for positioning. The algorithm was designed for use on a six-axis robot performing stationary tasks. As a result of the designed and tested system, the real-time positioning accuracy of the six-axis industrial robot greatly improved. Thus, the proposed algorithm can improve the positioning accuracy of robots in tasks, such as welding, riveting, and threading. The authors' vision for further research is to adapt this algorithm to other industrial robot manufacturers, such as ABB and Kuka [11].

In their experiment, Albert Nubiola et al. compared two metrology devices to calibrate an industrial robot. The metrology devices used were the FARO laser tracker and the Creaform C-Track. These calibrations were performed on an ABB IRB 120 industrial robot. The main difference between these devices is that the C-Track device works on the principle of photogrammetry. Through experimentation, the authors reduced the position error by almost 5 mm on average. At the same time, the authors pointed out the ease of use and the more affordable option of calibration with the Creaform C- Track device [12].

An experiment by Mohamed Slamani et al. investigated the positioning accuracy and repeatability of a six-axis ABB IRB 1600 industrial robot. The design of the test conditions was based on the ISO 9283 standard, which focuses on assessing the performance parameters of industrial robots. The authors used a trio of metrology devices: a laser tracker, a laser interferometer, and a Renishaw ballbar test. Using the laser tracker, the repeatability was measured in the range from 22 to 37 mm, respectively. The repeatability was measured between 6 and 14 mm using the Renishaw laser interferometer. The last device used was the circular ballbar test, with which a circularity deviation was measured within 1 mm. Of all the metrology devices used, the authors chose the circular ballbar test as the most suitable, as it is an accurate and affordable device that can measure industrial robots' static and dynamic parameters [13].

In a paper by Jerzy Józwik et al., they tested a repeatability of the Yaskawa Motoman HP20F industrial six-axis robot. They used a Phantom v2511 high-speed camera to measure the repeatability. The measurement principle was based on tracking the designed points and subsequent image analysis using the TEMA MOTION software. The measurement was performed in two planes, X and Y, respectively. The measured deviations were then added to the formulas according to ISO 9283. The result of this experiment was that as the number of cycles increased, the robot achieved a worsened repeatability. At the same time, the manufacturer's declared repeatability was RP = 0.06 mm, but the authors measured a repeatability that was several hundredths of a millimeter worse at several points [14].

In their experiment, the Michal Vocetka et al. investigated the effects of thermal expansion on the design of an industrial robot in order to reduce the difference in robot repeatability at low and high temperatures. The robot chosen for the experiment was an ABB IRB 1200 5/0.9. The authors proposed a compensation methodology to reduce the effect of drift on the robot's repeatability. A diagram incorporated in front of the robot control unit was able to evaluate and adjust the robot's position. As a result of the experiment, the robot's repetitive drift was reduced by almost 85% with the proposed compensation methodology. Prior to the introduction of the methodology, the repeatability was in the neighborhood of 200 μm; after the introduction, it was in the range of 25 μm at any temperature. Moreover, with the proposed methodology, the robot can be used immediately after startup, and there is no need to wait for the robot to warm up to the operating temperature [15].

3. Testing of Performance Characteristics According to ISO 9283

ISO 9283 describes the test methods for determining the performance characteristics of industrial robots. The standard contains 14 tests that can be used to develop, test, or compare industrial robots. The number of measurement points and cycles characterizes each test. The tests are described in detail to determine the correct measurement conditions to be observed. Each performance test must be performed at the maximum possible load and speed. The standard itself does not specify what metrology system should be used to measure the characteristics, which can be seen as a benefit as being more affordable, and accurate measurement systems can be used. The most common characteristics assessed are the unidirectional position accuracy (AP), unidirectional position repeatability (RP), path repeatability (RT), distance accuracy (AD), and distance repeatability (RD). Most manufacturers of industrial robots test to this standard. However, only positional repeatability is specified by manufacturers in the robot selection specifications. Most of the characteristics are tested using an imaginary ISO cube inserted into the robot's workspace so that it is where the robot most often operates [16–18].

In Figure 2, it is possible to see a cube on which a diagonal has been chosen to perform the tests using the points C1 to C8. In this example, these are points C1–C2–C7–C8. The next step is to select the measurement points, the number of which may vary depending on the characteristic being measured. In this example, the measured characteristic was unidirectional position accuracy. For the measurement, five measurement points (P1–P5) have been selected [16,17].

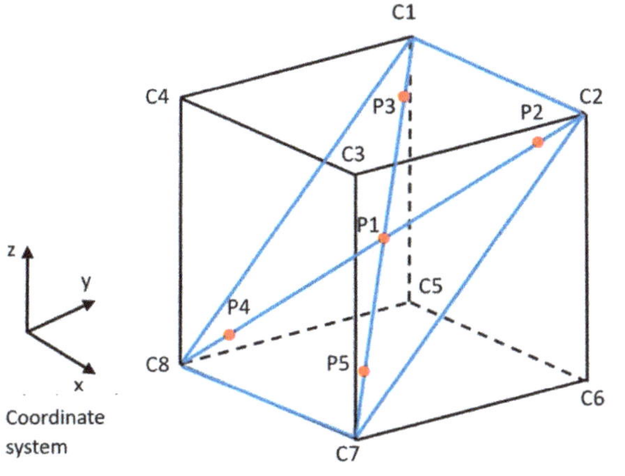

Figure 2. Example of an ISO cube.

1. Unidirectional position accuracy (AP): position accuracy is defined in the standard as the deviation between the desired position and the mean achieved value of the desired position (Figure 3). When performing the test, it is necessary to comply with the specified measurement conditions set out in the standard (Table 1): the mass of the maximum payload, the speed, and the number of repetitions [16–18].

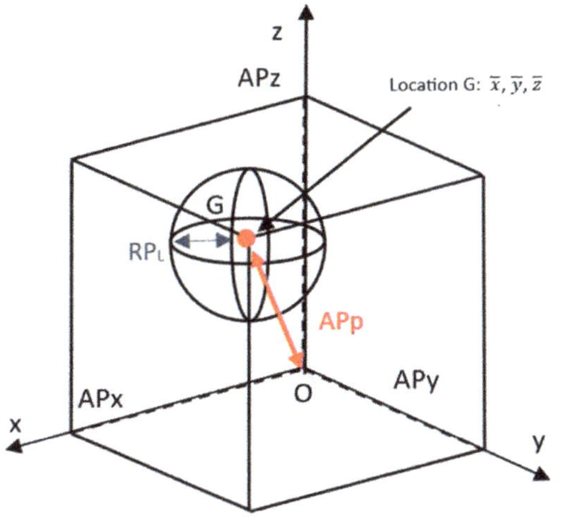

Figure 3. Position accuracy and repeatability.

Table 1. Position accuracy measurement conditions [16].

Performance Characteristics	Payload—% of Maximum Possible	Speed—% of Maximum Possible	Number of Measuring Points	Number of Cycles
Position accuracy	100%—required 10%—optional	100%—required 50%—10% optional	5 (P1–P5) 5 (P1–P5)	30 30

Unidirectional positioning precision (AP_p): the difference between the desired position and the range of positions achieved. The positioning accuracy is given by Equation (1):

$$AP_p = \sqrt{(\bar{x} - x_c)^2 + (\bar{y} - y_c)^2 + (\bar{z} - z_c)^2}$$
$$AP_P = \sqrt{(AP_x)^2 + (AP_y)^2 + (AP_z)^2} \quad (1)$$

where $\bar{x}$, $\bar{y}$, $\bar{z}$ represent the coordinates of the focal point obtained for a certain position after performing n number of cycles, and x_j, y_j, z_j (Equation (2)) represent the measured values. The terms x_c, y_c, and z_c represent the coordinates of the controlled, i.e., programmed, position [16].

$$\bar{x} = \frac{1}{n}\sum_{j=1}^{n} x_j\ ;\ \bar{y} = \frac{1}{n}\sum_{j=1}^{n} y_j;\ \bar{z} = \frac{1}{n}\sum_{j=1}^{n} z_j \quad (2)$$

2. Unidirectional position repeatability (RP): according to the above-mentioned standard, position repeatability is expressed as the difference between the positions reached after n cycles performed in the same direction for the same position [16,18]. For testing, the same conditions apply, as written in Table 1.

The unidirectional repeatability of the positioning is expressed by the RP_l value, and is calculated as shown in the following Equation (3):

$$RP_l = \bar{l} + 3S_l \quad (3)$$

where $\bar{l}$ represents the mean of the variable L, and S_l represents the standard deviation $\bar{x}$, $\bar{y}$, $\bar{z}$, and x_j, y_j, z_j is defined in Equation (4).

$$\bar{l} = \frac{1}{n}\sum_{j=1}^{n} l_j\ ;\ l_j = \sqrt{(x_j - \bar{x})^2 (y_j - \bar{y})^2 (z_j - \bar{z})^2}\ ;\ S_l = \sqrt{\frac{\sum_{j=1}^{n} (l_j - \bar{l})^2}{n-1}} \quad (4)$$

4. Methodology

4.1. Panasonic TM 2000 Industrial Robot

Panasonic manufactures and supplies several series of robots. The TM series of robots were designed primarily for welding processes. The mentioned welding robot series came with several improvements, such as the possibility of cable routing to the torch (internal and external) and stronger and more stable robot construction. The main improvement over the older types, where an absolute encoder was used on the servomotor, is using a high-precision smart encoder, which ensures a better track accuracy and more precise weld execution. The basic parameters of the used welding robot given by the manufacturer are presented in Figure 4.

The primary software for programming and controlling Panasonic robots is the DTPS software. This software has been exclusively designed for Panasonic industrial robots for offline programming. In this software, it is possible to create an overall welding design with cycle counts and welding times and simulate the manufacturing process. The advantage of these simulation programs is the prevention of possible collisions with the robotic equipment. This means that when directly programming from the CAD model, the worker can change the welding torch's orientation to avoid unwanted robot positions. The software supports multiple formats for CAD model insertion (STL, IGES, IGS, and DXF). Panasonic has also incorporated robot and external positioner axis calibration into the software.

Model	TM – 2000
Maximum payload	6 kg
Maximum reach	1 999 mm
Minimum reach	491 mm
Number of axes	6
Maximum speed	180 m/min
Position repeatability	± 0.10 mm
Maximum movement speed	RT = 195°/s
	UA = 197°/s
	FA = 205°/s
	RW = 425°/s
	BW = 425°/s
	TW = 625°/s
Weight	216 kg

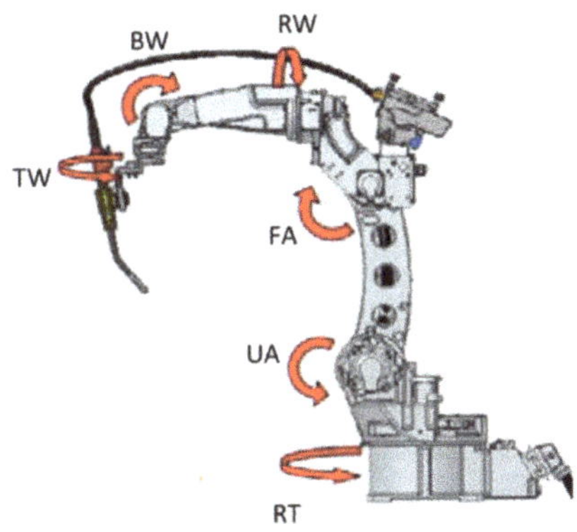

Figure 4. Panasonic TM-2000 welding robot and its basic characteristics.

4.2. Design of Measuring Equipment, ISO Cubes, and the Selection of the Test Conditions

A digital strain gauge from INSIZE, model 2112-254F, was used to implement the column measurements. It is an accurate, simple, and affordable measuring device. It has an accuracy of 0.005 mm, a measuring range of 25.4 mm, and a digital step of 0.001 mm for digital measurements, potentially storing the data in the form of a table of values directly on the PC. As the repeatability was measured on a welding robot having a welding wire at the end of the welding nozzle, a 20 mm diameter contact area was chosen for the repeatability measurement due to the impossibility of attaching another suitable object, e.g., a precision machined cube, to the end of the robotic arm. A fixture was subsequently designed and fabricated to fix the measuring device to the welding table. The complete design of the measuring station in real conditions in the production hall is shown in Figure 5.

Figure 5. Measuring system applied to a robotic workstation.

Once the measurement system was selected, designing an ISO cube with a selected measurement plane and five measurement points was the next necessary step. The dimensions of the ISO cube were determined based on the robot's workspace and where the robot most often welds. The length of the edge was 1200 mm. This type of welding robot uses the "touch sensing" function, which means that the robot locates the position of the workpiece to be welded before welding. This type of smoothing is entered directly when programming the welding task of the robot. It works on the principle of the tension generated when the welding wire comes into contact with the material to be welded. In this way, it is possible to prevent inaccurate welds, which may have been caused, for example, by the displacement of the workpiece or inadequate calibration of the positioner.

Due to the "touch sensing" function, the Z-axis was omitted from the measurement as the robot can determine the distance from the workpiece in the Z-axis when welding. The proposed ISO cube is displayed in Figure 6, where the points C5–C6–C7–C8 represented the selected measurement plane and P1 to P5 represented the measured points, respectively.

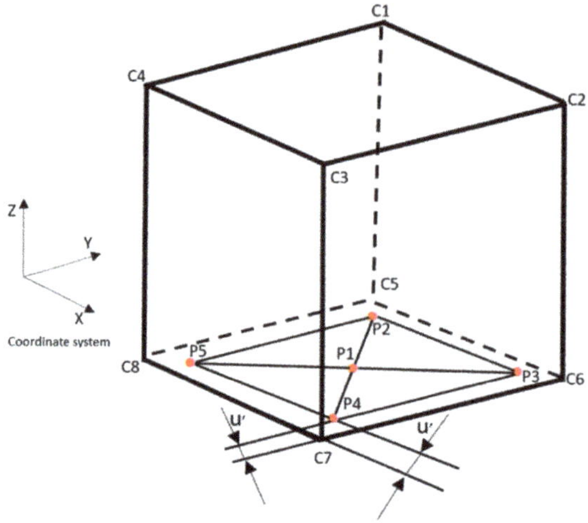

Figure 6. The proposed ISO cube with measurement points.

Commonly used welding speeds range from 30 to 60 m/min, respectively. For this reason, the speed in the measurement was set at 90 m/min, representing 50% of the maximum possible speed. Other test conditions are shown in Table 2.

Table 2. Determination of the test conditions.

Movement Speed 90 m/min	Load 3 kg	Tested Positions	Number of Performed Cycles	Measured Axis
50% of the maximum possible	50% of the maximum possible	P1-P2-P3-P4-P5	30 Cycles	X–Y

5. Creation and Design of a Simulation of a Robotic Welding Workstation

RoboDK software was chosen for the design of the workstation simulation. It is a simulation software that offers an intuitive environment in which various robotic tasks can be programmed, such as welding, palletizing, packaging, milling, 3D printing, etc. In addition, the software supports basic CAD model input formats, such as STEP, IGES, and STL. The advantages of this software compared to other software are as follows [19]:

- An online library containing a large number of robots, positioners, and tools from various manufacturers;
- The possibility to insert your own designed robot in the CAD program;
- The possibility of testing the precision of robots;
- The software runs performance tests according to ISO 9283 (Renishaw ballbar test), and the calibration of robots using laser trackers;
- Exporting post processors for the different types of controllers.

This simulation was possible thanks to creating CAD components of the welding station according to the layout of the real workstation. First, the dimensions of the workstation, the positioner, the welding table, and the robot travel were measured. The individual functional units of the equipment are shown in Figure 7.

Figure 7. Robot workstation in the production hall.

After measuring the main components, it was necessary to determine the layout of the components in the space. After obtaining the dimensions, the individual parts of the station were processed into CAD models using Creo Parametric software. The basic step in creating the station was the insertion and appropriate positioning of the 4500 × 3000 mm floor, which was available from the online library. The next step was inserting the pre-prepared CAD models of the workstation and their layout. In order to insert the files into the software environment, it was necessary to save the components in STEP format, which RoboDK supports. First, barriers with a back wall were added to form the boundary of the workstation. This was followed by inserting the welding station's most important parts: the positioner, and the vehicle with the base. The Panasonic TM-2000 robot was downloaded from the online library and added to the assembly, where it was placed onto the base with the traveler. Other components added to complete the welding station were a welding table, pallet truck, control box, welding wire, positioner switch, teach pendant, guard markings, and weld parts. The complete design can be seen in the RoboDK software simulation environment, which was as shown in Figure 8.

Figure 8. Workstation layout in the simulation model setting.

Programming and Adding Targets for the Robot

After creating the simulation model of the welding station, the robot motion sequences simulated for the measurement were created. Next, the designed ISO cube with an edge length of 1200 mm was inserted into the robot workspace, which is shown in Figure 9. The cube was inserted at the place where the robot welds most frequently, which was one of the conditions for measuring the repeatability and positioning accuracy of the robot arm.

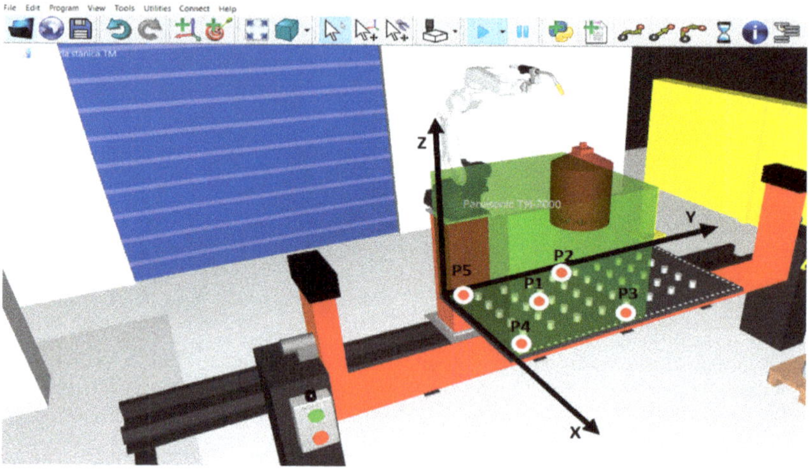

Figure 9. ISO cube with the designed measurement points in the robot's workspace.

Programming the robot motion sequences in RoboDK and thus preparing the measurement program consists of assigning the targets to be reached by the robot. In this case, the targets for the robot were added using the "Teach target(s) on surface" function, located in the "Program" table. By selecting this function, the TCP robot moved in parallel with the selected target by fixing it to a well-defined point on the surface of the inserted cube. In this case, it was the contact area where the contact of the deflector with the welding head occurs. Since the digital indicator has a measuring range of 25.4 mm, the robot needed to compress the contact by 10 mm. After creating all the targets for the robot, the next step was to add and generate the program. In this case, two programs were created, one for the X-axis and one for the Y-axis, respectively, for better transparency.

6. Measurement and Evaluation of Results

During the measurements, it was necessary to observe further rules, which are set out in ISO 9283, such as:

- The robot must be mounted under the manufacturer's control and fully functional during the measurement;
- The ambient temperature must be in the range 20–22 °C;
- Both the robot and the measuring device must be in a thermally stable condition, and must not be affected by scientific influences, e.g., sunlight;
- Before measuring the selected characteristic, the robot must undergo a warm-up cycle, including the drift measurement.

As it was necessary to observe the warm-up cycle, the measurements were made after an 8 h working shift under optimal working conditions: pressure 101.3 kPa, temperature 21 °C, and humidity 38%. After this time, the robot was sufficiently warmed up and ready for measurement. The digital aberration meter was placed in the robot area a day before the measurements were taken to ensure temperature stability.

The actual measurement was preceded by placing the deflection meter on the welding plate. The next step was to run the program on the robot in the so-called test mode, in which the speed was significantly reduced. This mode was used for every newly created program to assess the control program's correctness.

The robot arm position repeatability was measured after testing the program in the test mode. First, the digital indicator was fixed to the desired position. Next, a cable for data transmission was connected to the digital indicator and a notebook with a blank MS Excel workbook, where the measured values were automatically saved. At the first occurrence of a touch on the digital indicator, the digital indicator was reset at all measurement points so that the reading could be taken directly from the selected point. When the robot entered the programmed position, a 5 s pause allowed the value on the deflector to settle, and then the value was written to the PC by pressing a button on the cable. Figure 10 shows the aberration meter at the measured point P1 on the X-axis and Y-axis.

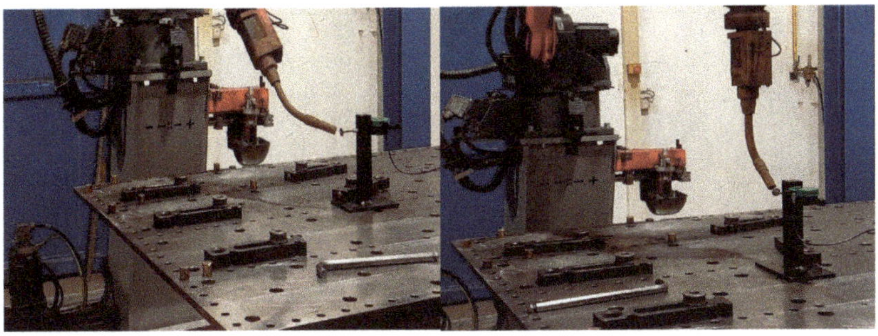

Figure 10. X-axis and Y-axis measurement at P1.

The X-axis was measured first, and the measurement started at P1. After 30 readings were recorded for point P1, the program was paused, and the jig with the digital indicator was moved to position P2. The measurement continued similarly at points P3, P4, and P5. Table 3 shows the measured X-axis deviation values.

Table 3. Measured X-axis deviation values.

Measurement Number	Measured Points in the X-Axis				
	Point P1	Point P2	Point P3	Point P4	Point P5
1	0.077	−0.033	−0.005	−0.018	−0.002
2	0.091	−0.013	−0.009	−0.028	0.003
3	0.088	0.019	−0.014	−0.035	−0.027
4	0.087	0.011	−0.008	−0.033	0.011
5	0.122	0.007	−0.012	−0.040	0.016
6	0.101	0.002	−0.021	−0.038	0.011
7	0.101	−0.016	−0.029	−0.040	0.025
8	0.121	−0.011	−0.024	−0.032	0.022
9	0.122	−0.003	−0.028	−0.029	0.028
10.	0.108	0.002	−0.015	−0.031	0.022
11	0.110	0.018	−0.033	−0.039	0.020
12	0.112	0.006	−0.016	−0.044	0.012
13	0.104	0.004	−0.004	−0.031	−0.010
14	0.132	0	−0.007	−0.035	−0.007
15	0.112	−0.001	−0.008	−0.039	−0.017
16	0.066	−0.009	−0.013	−0.037	−0.025
17	0.149	−0.008	0.015	−0.038	−0.026
18	0.059	−0.001	0.001	−0.049	0.016
19	0.133	−0.011	0.001	−0.043	0.016

Table 3. *Cont.*

Measurement Number	Measured Points in the X-Axis				
	Point P1	Point P2	Point P3	Point P4	Point P5
20	0.065	−0.001	−0.001	−0.046	−0.012
21	0.062	−0.005	−0.027	−0.037	0.008
22	0.122	−0.004	−0.029	−0.045	−0.024
23	0.063	0	−0.014	−0.045	−0.002
24	0.053	0.006	−0.033	−0.044	0.016
25	0.056	−0.003	−0.031	−0.043	−0.021
26	0.132	−0.001	−0.020	−0.051	0.049
27	0.110	0.001	−0.029	−0.057	0.038
28	0.107	0.006	−0.029	−0.058	−0.013
29	0.059	−0.006	−0.033	−0.056	−0.020
30	0.048	0	−0.012	−0.049	0.024

The measurement continued with the control program for the Y-axis, where the jig with the deflection gauge was rotated by 90° and set to position P1. On the first touch, the digital indicator was reset, and then on the next 30 touches, the values were recorded. As with the X-axis point measurements, the measuring device was moved to the remaining positions (P2, P3, P4, and P5) where 30 values were recorded for each point, as summarized in Table 4.

Table 4. Measured Y-axis deviation values.

Measurement Number	Measured Points in the Y-Axis				
	Point P1	Point P2	Point P3	Point P4	Point P5
1	−0.017	0.009	−0.009	−0.009	0.018
2	−0.007	0.010	0.003	0	0.008
3	−0.022	0.009	0.006	−0.013	0.005
4	−0.019	0.016	−0.006	0.007	0.010
5	0	−0.003	−0.012	0.005	0
6	−0.014	0.013	0.045	0.007	0
7	0.004	0.004	0.034	0	−0.002
8	−0.021	0.005	0.016	−0.013	0.001
9	0.011	−0.007	0.032	−0.002	−0.009
10	0.026	−0.015	−0.005	0.007	0.012
11	−0.016	0	0.002	−0.001	−0.005
12	−0.037	−0.005	−0.015	0.016	0
13	−0.004	−0.004	0.022	0.001	−0.015
14	−0.012	−0.001	0.008	0.015	−0.015
15	−0.038	−0.018	0.036	−0.008	0.007
16	−0.024	0	0.046	−0.002	−0.007
17	−0.018	−0.009	0.020	0.002	0.008
18	−0.026	−0.015	0.035	−0.004	−0.018
19	−0.027	−0.002	0.026	0.006	−0.002
20	−0.045	−0.004	0.031	0	0.023
21	−0.029	−0.005	0.036	0.004	0
22	−0.037	−0.028	0.030	−0.007	−0.001
23	−0.036	−0.015	0.033	0.004	−0.003
24	−0.051	−0.006	0.033	−0.007	0.011
25	−0.043	−0.008	0.038	−0.007	0.018
26	−0.057	−0.002	0.033	−0.004	0.014
27	−0.048	−0.015	0.014	0.005	0.026
28	−0.041	−0.002	0.019	0.005	−0.004
29	−0.036	−0.011	0.010	0	0.001
30	−0.131	−0.001	0.033	0.024	−0.013

7. Discussion

A total of 300 measurements were obtained, followed with a repeatability count. The repeatability was calculated for each measurement point separately. From the measured values, it was impossible to calculate the average one-way repeatability of the position as the equipment was moved when measuring the deviations of one point. Since the measurement was made in the X and Y axes, the axes were measured separately.

In particular, using the formula to calculate the total unidirectional position repeatability as specified in the standard was not possible. For this reason, an equation based on the standard deviation (Equation (5)), where the index p_n denotes the point to be measured, was used to calculate the repeatability.

$$RP_{Pn} = \pm 3\, S_{Pn} = \pm 3 \sqrt{\frac{\sum_{j=1}^{n} \left(P_{nj} - \bar{P}n\right)^2}{n-1}} \quad (5)$$

Subsequently, the results obtained from the calculation were compared with the data reported by the manufacturer, which are presented in Table 5. The measurement results for the X-axis clearly show that the robot achieved an even better repeating in the X-axis than that given by the manufacturer. Furthermore, comparing the results from the Y-axis measurement with the value given by the manufacturer also showed that the robot achieved a better repeatability.

Table 5. Comparison of the calculated X- and Y-axis values.

Measured Point	Repeatability in the X-Axis	Repeatability in the Y-Axis
Point P1	±0.086 < 0.1 mm	±0.082 < 0.1 mm
Point P2	±0.030 < 0.1 mm	±0.029 < 0.1 mm
Point P3	±0.037 < 0.1 mm	±0.053 < 0.1 mm
Point P4	±0.027 < 0.1 mm	±0.025 < 0.1 mm
Point P5	±0.062 < 0.1 mm	±0.033 < 0.1 mm

It was also possible to examine the graphical processing of the calculated standard deviations for the points on the X-axis, as shown in Figure 11.

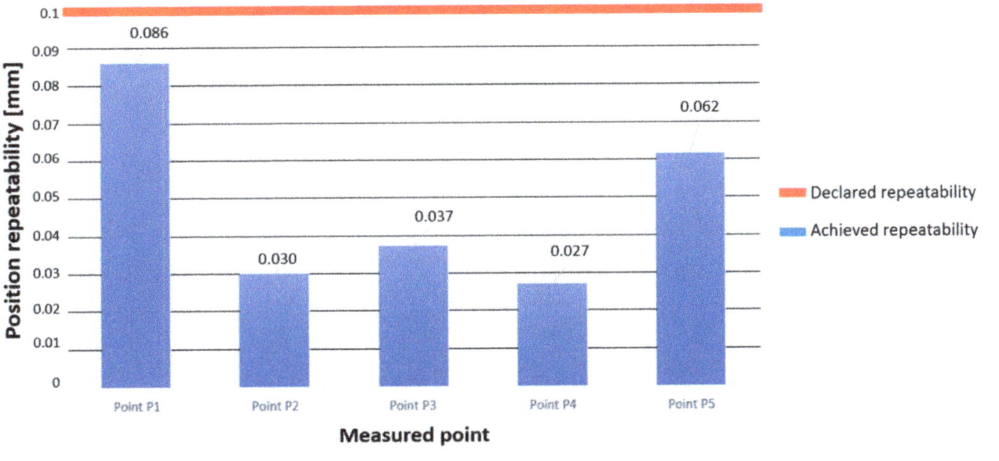

Figure 11. Position repeater for points in the X-axis.

The graph for the Y-axis was also processed in the same way, as shown in Figure 12.

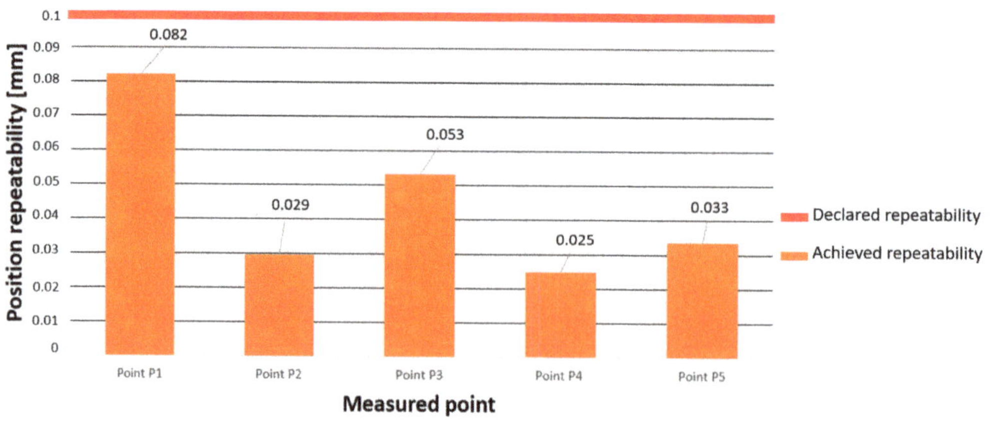

Figure 12. Position repeater for points in the Y-axis.

From graphically comparing the achieved values in the X and Y axes, it is shown in in Figure 13 how the robot achieved an almost identical repeatability values at points P1, P2, and P4. The assumption was that at points P3 and P4, which were at the boundary of the working space and thus when the robot's arm was the most extended, it would achieve the worst repeatability, but this assumption was not confirmed. The worst repeatability for both measurements was achieved at point P1, which was located in the middle of the designed measurement plane.

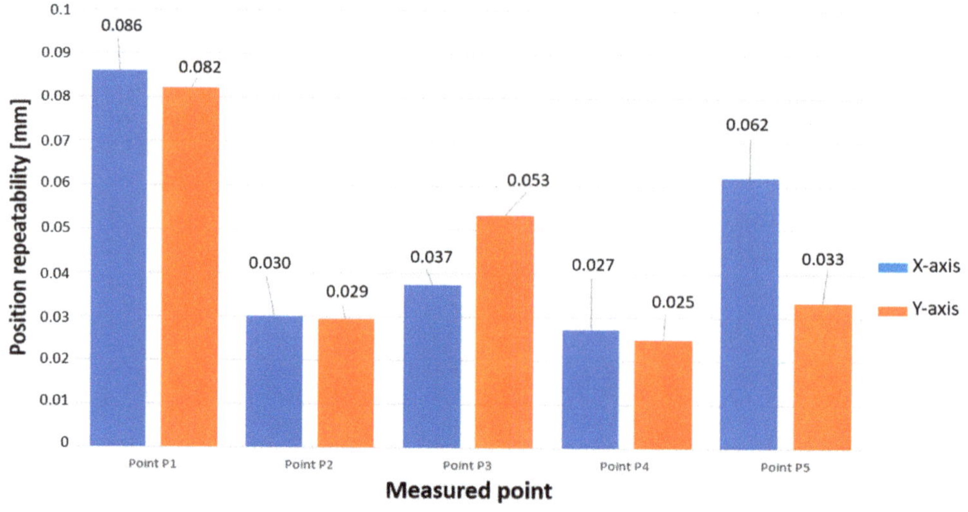

Figure 13. Comparison of the results of the measured values.

The specific result we obtained was that the robot achieved a better repeatability at each point than the manufacturer indicated. However, the measured values at point P1 indicated that the repeatability around the measured point may deteriorate.

8. Conclusions

The ISO 9283 standard forms the basis for assessing the performance characteristics of industrial robots. The unidirectional position repeatability is one of the most important parameters to be monitored regarding the reported accuracy of robots. Most manufacturers declare the repeatability and position accuracy based on tests they performed according to the ISO 9283 standard. However, in the case of Panasonic, it could not be clearly established on how and on what basis the manufacturer tests its industrial robots. The following factors contribute to the deterioration of position repeatability: irregular calibration of the robot axes, arm overstressing, or collisions after which the robot has not been calibrated and inspected.

The object of study described in the paper was a Panasonic TM-2000 welding robot, which welds the end frames for forklift trucks in the factory. This paper discussed the selection of a suitable metrological device for a quick repeatability check and a possible calibration of the robot directly at the workplace, with an emphasis on the simplicity and efficiency of the device's measurement for workers working at a welding robotic workplace. The RoboDK software was chosen for the simulation design of the trajectories and the generation of the control program for the realization of the robot repeatability measurements. The measurements and the simulations were possible thanks to designing and manufacturing a measuring device, an imaginary ISO cube, a measuring plane, measuring points, and a cycle to determine in which order the points would be measured. This paper further described the design procedure of the ISO cube, which formed the basis for the design of the measurement plane and the measurement points. A digital indicator was chosen as the measuring device on which a flat touch was used. The disadvantage of the digital indicator is the impossibility of measuring in the Z-axis, but in this case, the measurement was made in the X and Y axes. As the robot included a touch sensing function, it was not necessary to perform measurements in the Z-axis as specified by the ISO 9283 standard as the robot verified the position of the welding nozzle through an integrated sensor during calibration before starting the welding process.

The analytical part resulted in a direct measurement of the position repeatability on the welding robot. A total of five points were measured for the X-axis and five points for the Y-axis, respectively. Each point was recorded 30 times, and the measurements were taken in the positive direction of motion. These results were compared with the value provided by the manufacturer, and the measured deviations were also processed graphically. The robot achieved better results compared the manufacturer's results at each point. After comparing the results in both axes, the trend in that the more the robot's arm is stretched, the worse the repeating the robot achieves, was not confirmed. The assumption we made was that at point P1, where the robot has achieved the worst repeatability, the repeatability and accuracy may deteriorate in the future.

After consulting the results with the quality department of the manufacturing company, the following recommendations were proposed for maintaining the quality and repeatability of the robot:

- Regularly perform calibration of all robot axes and also their positioners;
- Regularly check the shape of the welding torch and correct any irregularities;
- Using the designed measuring device to check the repeatability of the robot on the trajectories of the busiest workplace points;
- In the event of a robot collision, perform calibration, and then check the repeatability with the designed measuring device.

Author Contributions: Conceptualization, K.G. and M.P.; methodology, M.P.; software, K.G.; validation, M.P.; formal analysis, K.G. and M.P.; investigation, M.P.; resources, K.G.; data curation, K.G.; writing—original draft preparation, K.G. and M.P.; writing—review and editing, M.P.; visualization, K.G.; supervision, M.P.; project administration, M.P. All authors have read and agreed to the published version of the manuscript.

Funding: This research received no external funding.

Data Availability Statement: Not applicable.

Acknowledgments: The authors thank the Ministry of education of Slovak Republic for supporting this research by the grant VEGA no. 1/0121/23, KEGA no. 038TUKE-4/2021, project APVV-18-0316 and APVV-19-0550.

Conflicts of Interest: The authors declare no conflict of interest.

References

1. Muelaner, J.E.; Wang, Z.; Maropoulos, P.G. Concepts for and Analysis of a High Accuracy and High Capacity (HAHC) Aerospace Robot. *Proc. Inst. Mech. Eng.* **2011**, *225*, 1393–1399. [CrossRef]
2. Niku, S.B. *Introduction to Robotics: Analysis, Control, Applications*, 3rd ed.; Standards Information Network: Chichester, UK, 2019; ISBN 9781119527602.
3. Placyek, M.; Pisycyek, L. Testing of an Industrial Robot's Accuracy and Repeatability in Off and Online Environment. *Eksploat Niezawodn-Maint. Reliab.* **2018**, *20*, 455–464. [CrossRef]
4. Broum, T.; Simon, M. Safety Requirements Related to Collaborative Robots in the Czech Republic. *MM Sci. J.* **2020**, *1*, 3852–3856. [CrossRef]
5. Hu, M.; Wang, H.; Pan, X.; Tian, Y. Optimal Synthesis of Pose Repeatability for Collaborative Robots Based on the ISO 9283 Standard. *Ind. Rob.* **2019**, *46*, 812–818. [CrossRef]
6. Motta, J.M.; Carvalho, G.C.; McMaster, R.S. Robot Calibration Using a 3D Vision-Based Measurement System with a Single Camera. *Robot. Comput. Integr. Manuf.* **2001**, *17*, 487–497. [CrossRef]
7. Ondocko, S.; Svetlik, J.; Sasala, M.; Bobovsky, Z.; Stejskal, T.; Dobransky, J.; Demec, P.; Hrivniak, L. Inverse Kinematics Data Adaptation to Non-standard Modular Robotic Arm Consisting of Unique Rotational Modules. *Appl. Sci.* **2021**, *11*, 1203. [CrossRef]
8. Zaborowski, T.; Panda, A.; Androvic, L. Robots and Cobots, Main Differences. *Stud. Mater.* **2019**, *39*, 4–8.
9. Landstorfer, P.; Hiller, J.; Herbst, M. Investigation of Positioning Accuracy of Industrial Robots for Robotic Based X-ray Computer Tomography. In Proceedings of the 9th Conference on Industrial Computed Tomography, (iCT 2019), Padova, Italy, 13–15 February 2019; pp. 1–5. [CrossRef]
10. Stejskal, T.; Svetlik, J.; Ondocko, S. Mapping Robot Singularities through the Monte Carlo Method. *Appl. Sci.* **2022**, *12*, 8330. [CrossRef]
11. Gharaaty, S.; Shu, T.; Joubair, A.; Xie, W.F.; Bonev, I.A. Online Pose Correction of an Industrial Robot Using an Optical Coordinate Measure Machine System. *Int. J. Adv. Robot. Syst.* **2018**, *15*, 1729881418787915. [CrossRef]
12. Nubiola, A.; Slamani, M.; Joubair, A.; Bonev, I.A. Comparison of Two Calibration Methods for a Small Industrial Robot Based on an Optical CMM and a Laser Tracker. *Robotica* **2014**, *32*, 447–466. [CrossRef]
13. Slamani, M.; Nubiola, A.; Bonev, I. Assessment of the Positioning Performance of an Industrial Robot. *Ind. Robot.* **2012**, *39*, 57–68. [CrossRef]
14. Józwik, J.; Ostrowski, D.; Jarosz, P.; Mika, D. Industrial Robot Repeatability Testing with High Speed Camera Phantom V2511. *Adv. Sci. Technol. Res. J.* **2016**, *10*, 86–96. [CrossRef]
15. Vocetka, M.; Bobovský, Z.; Babjak, J.; Suder, J.; Grushko, S.; Mlotek, J.; Krys, V.; Hagara, M. Influence of Drift on Robot Repeatability and Its Compensation. *Appl. Sci.* **2021**, *11*, 10813. [CrossRef]
16. *ISO 9283:1998*; Manipulating Industrial Robots—Performance Criteria and Related Test Methods. ISO: Geneva, Switzerland, 1998.
17. Józwik, J.; Jacniacka, E.; Ostrowski, D. Uncertainty Measurement with the Kinematic Telescopic Bar during Industrial Robot Inaccuracy Tests. *ITM Web Conf.* **2017**, *15*, 04013. [CrossRef]
18. Kuric, I.; Tlach, V.; Ságová, Z.; Císar, M.; Gritsuk, I. Measurement of Industrial Robot Pose Repeatability. *MATEC Web Conf.* **2018**, *244*, 01015. [CrossRef]
19. Simulator for Industrial Robots and Offline Programming—RoboDK. Available online: https://robodk.com/ (accessed on 29 May 2023).

machines

Article

Analysis of the Design of the Single-Cylinder Steam Engine of the Grasshopper Beam by Henry Muncaster

José Ignacio Rojas-Sola [1,*] and José Francisco Gutiérrez-Antúnez [2]

1 Department of Engineering Graphics, Design and Projects, University of Jaen, 23071 Jaen, Spain
2 Higher Polytechnic School, University of Jaen, 23071 Jaen, Spain
* Correspondence: jirojas@ujaen.es; Tel.: +34-953-212452

Abstract: In this paper, the analysis of the design of the single-cylinder steam engine of the Grasshopper beam designed by Henry Muncaster in 1912 is shown. This engine was incorporated into ships and railways and it was published in the *Model Engineer* journal in 1957. It shows great complexity due to its ability to transform reciprocating movement into rotary movement and its high number of elements (more than 150). To this end, a study of computer-aided engineering (CAE) was carried out using the software Autodesk Inventor Nastran, consisting of a linear static analysis using the finite element method (FEM) of the 3D CAD model under real operating conditions in order to learn if it was well designed, according to material resistance criteria. The assembly is analyzed in the two most unfavorable situations in order to determine the von Mises stresses, the displacements and the safety coefficient distributions. Finally, it was found that the work pressure (maximum admissible steam pressure in the admission) was 0.15 MPa.

Keywords: Henry Muncaster; steam engine; grasshopper beam; Autodesk Inventor Nastran; computer-aided engineering; finite-element analysis

1. Introduction

Steam engines have historically been one of the most widely used inventions in engineering, mainly between the 18th and 19th centuries when steam was the most commonly used form of energy. In reality, they are constituted as thermal machines, where the expansion of the steam occurs, transforming the heat energy into mechanical energy (work), whose magnitude will depend on the temperature gradient between the temperature of the steam at the entrance and the temperature of steam at the outlet of the machine [1,2].

Among the first to put this machine into service were Papin, Savory and Newcomen; Humphrey Potter, who originated the automatic operation of the valves; James Watt [3], who invented the separate condenser; and Trevithick, Hedley and Hackworth, the pioneers of the locomotive. Thus, it is a machine that has been used for different purposes and implemented in different means of locomotion, such as ships or railways, and that has undergone modifications to improve its efficiency and economy.

Henry Muncaster was an English engineer who was born in the second half of the 19th century, an expert in everything related to steam engines and who developed his work in the mining, coal, steel and large pumping industries [4]. From the beginning of the 20th century, he contributed numerous articles to the *Model Engineer* magazine until his death in 1935, and in 1912, he published a book presenting eight stationary steam-powered devices [5]. The present investigation analyses one of his inventions, the Grasshopper beam steam engine, from an engineering viewpoint [6,7].

Recently, an investigation carried out by the authors allowed us to obtain the 3D CAD model of this invention [8] from the plans of Julius de Waal [7], in which the operation of the device is explained in detail and on which this analysis is based.

The grasshopper beam steam engine pivots at one end rather than the center, and usually, the connecting rod of the crankshaft is positioned between the piston and the pivot, i.e., they act as a lever of the second kind [9]. In this way, a long stroke is produced for the piston with a shorter stroke for the crank, albeit with greater force.

The first recorded example was William Murdoch's model steam carriage of 1784 [10], although later its use was extended mainly to marine engines. Unlike conventional beam engines that required masonry construction to house the heavy beam and pivot, the grasshopper beam was pivoted on a swinging link. Thus, they were much lighter and their use spread to smaller engines [6]. Furthermore, although the weight of the Grasshopper beam is completely unbalanced, the fact that it is much lighter than the conventional beam allows Grasshopper engines to perform satisfactorily at higher speeds than conventional beam engines [7].

Moreover, there is no worldwide study of this engine from the point of view of engineering, which underlines the novelty, originality and scientific interest of this work. For this reason, the ultimate goal of this study was to perform a static analysis [11] of the single-cylinder steam engine of the grasshopper beam using the finite element method [12] under real operating conditions in order to determine whether it was properly designed and show how the model could be optimized by determining the work pressure (maximum admissible steam pressure in the admission).

The remainder of the paper is structured as follows: Section 2 shows the materials and methods (including the machine's operation) used in this research, Section 3 includes the main results (modal analysis and static analysis) and discussion, and Section 4 presents the main conclusions.

2. Materials and Methods

The starting material for this research was based on the 3D CAD model, obtained previously [8] using the Autodesk Inventor Professional software [13], to carry out a computer-aided engineering study, particularly a linear static analysis, using Autodesk Inventor Nastran [14]. This is a powerful software for the numerical simulation of engineering problems; particularly, it is a calculation engine for the finite element method that is well recognized in mechanics.

This section briefly explains the operation of the steam engine under study, in order to provide a better conceptualization of the analysis carried out and the working hypotheses from the design point of view, detailing each of the previous phases before executing linear static analysis. In this analysis, two of the most notable variables are the von Mises stress and the displacements suffered by the components of the machine, since these variables mechanically characterize the behavior of the assembly, verifying that its operation under real operating conditions is completely safe.

2.1. Operation of the Machine

Although the detailed operation of the machine is explained in detail in a recent investigation carried out by the authors [8], where detailed information on the modeling and assembly process can be obtained (Figure 1), a brief summary is presented below.

This machine is made up of a single double-acting cylinder. Therefore, the working fluid (water vapor) enters and leaves on both sides of the piston, thus carrying out two thermodynamic cycles simultaneously (although out of phase; for example, while on one side of the piston, the admission of water vapor occurs, on the other, the escape occurs). Intake and exhaust control are carried out through a slide valve inside the slide valve chest block, which is located next to the piston cylinder. The movement of this slide valve is due to a linkage, whose relative movement between parts is produced thanks to the coupling of the connecting rod of the slide valve to the crankshaft. An eccentric sheave is attached to it in order to synchronize the movement of the piston with the movement of the slide valve so that the admission and exhaust of the steam are carried out during certain periods of time, thus achieving good operation of the machine.

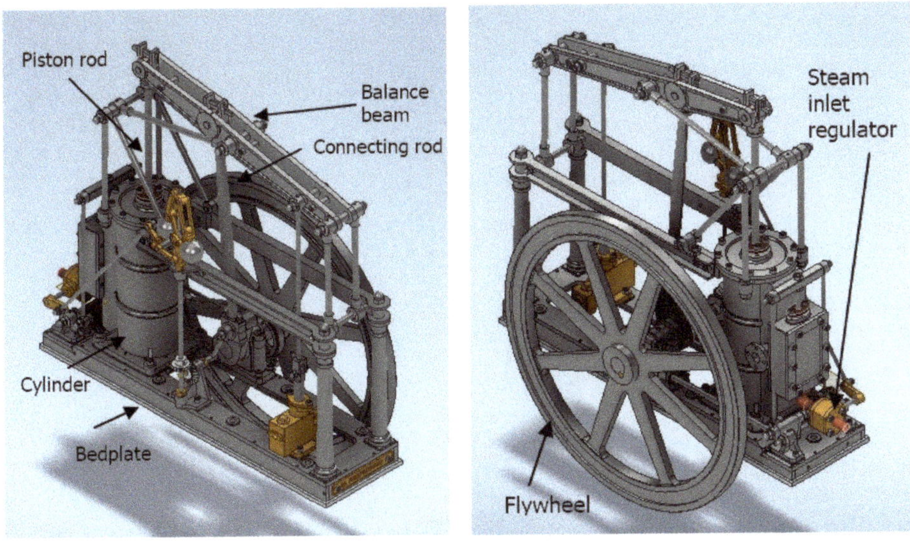

Figure 1. Assembly of the machine: front view (**left**) and rear view (**right**).

A notable element of the machine and its design is the master linkage used to transform the reciprocating movement of the piston into a rotary movement of the crankshaft (Figure 2). For this, a balance beam is used, located in the upper part of the invention, which allows a lightening of the weight of the machine.

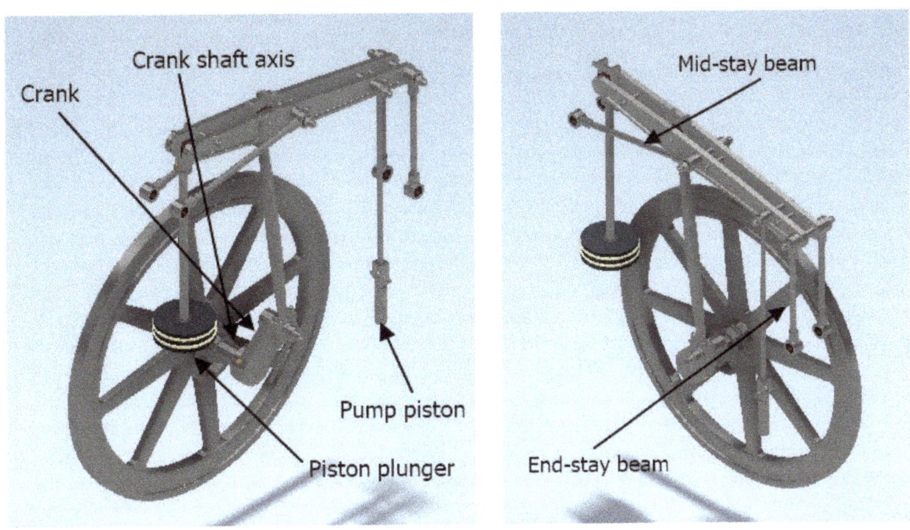

Figure 2. Master linkage: front view (**left**) and rear view (**right**).

In addition, a connecting rod with a piston at its end is attached to the balance beam, which will have an alternative movement inside a pump to compress the water vapor and propel it from the tank to the boiler, from where it will come out with higher pressure and temperature, to later expand in the cylinder chamber and transform the thermal energy into mechanical energy, which will be used thanks to the flywheel.

A great advantage of this machine, in addition to the one mentioned above, is its ability to self-regulate and keep working under stationary conditions. For this, a speed regulator is used (Figure 3) with a very specific function: to regulate the flow of steam inlet to the slide valve chest block by turning a valve that will increase or decrease the cross-section inside the steam inlet housing. As a consequence, if the velocity of the fluid through the cross-section changes, the flow will also change. In this invention, the rotation of this valve is carried out automatically, being controlled by the speed regulator.

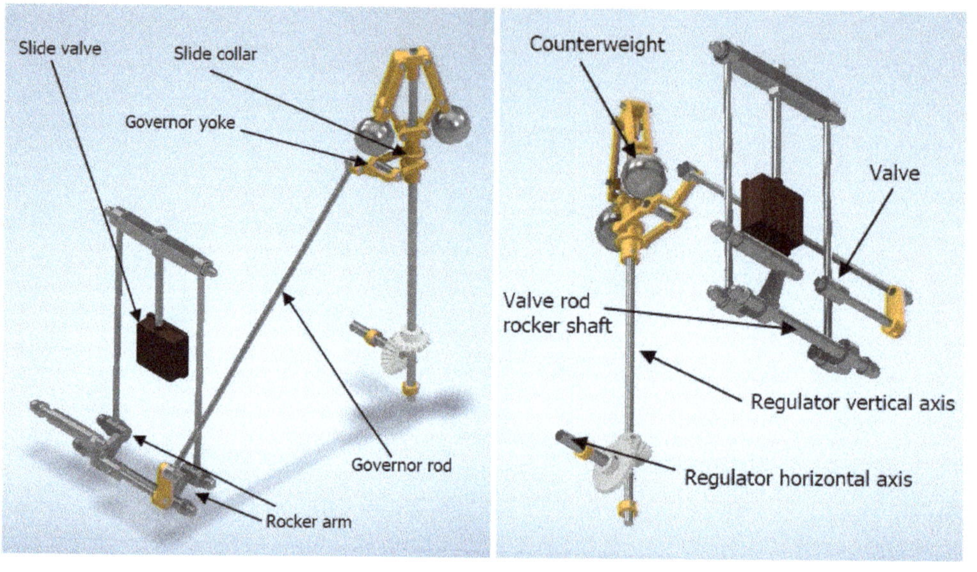

Figure 3. Speed regulator: top view (**left**) and bottom view (**right**).

To accomplish this, the speed regulator is made up of two axes, one horizontal and the other vertical, each with its respective components. The horizontal axle is coupled to the crankshaft, and at its end is a gear that transmits the rotational movement from the horizontal axle to the vertical axle. At the opposite end of the vertical axle gear, there is a small linkage, formed by two articulated quadrilaterals, with some counterweights at the end of one of its links. Since the shaft rotates, these counterweights produce a centrifugal force that depends on the speed of rotation of the shaft and its angular acceleration. This causes the sliding collar of the speed regulator to move vertically, moving a yoke which in turn is connected to a bar that will rotate the valve connecting rod, producing its rotation and varying the inlet flow. Therefore, depending on the speed of the crankshaft, the centrifugal force exerted will vary and, consequently, the position of the sliding collar, thus regulating the inlet flow of water vapor. In this way, the speed of rotation of the flywheel is always the same and the steam engine works under stationary conditions since these steam engines were installed in trains or boats, with the speed constant through almost the entire displacement.

Figure 4 shows an axonometric view and a front view of the sectioned invention, in order to more clearly observe the elements that make up the machine, as well as the interior of the cylinder chamber and the slide valve chest block.

Figure 4. Sectioned views of the assembly: axonometric view (**left**) and front view (**right**).

2.2. *Analysis from the Mechanical Engineering Point of View*

This section explains all the stages of the process followed to carry out an analysis using the finite element method, as well as all the established working hypotheses. The stages are as follows:

- Pre-processing;
- Assignment of materials;
- Application of contacts;
- Boundary conditions;
- Discretization.

Subsequently, and once the previous stages have been carried out, a modal analysis was performed to determine the natural vibration frequencies of the machine, as well as a linear static analysis in the two critical positions of the piston plunger (lower dead center and upper dead center) which constitute the two critical positions analyzed.

2.2.1. Pre-Processing

An analysis could be carried out that included the steam engine as a whole, but due to the high number of components (more than 150) that make it up, the computational cost would be very high, and so would the simulation time. For this reason, elements that will not significantly influence the results obtained have been excluded from the analysis, given that their operating conditions will be far from critical (for safety factors regarding the unity). These excluded components are:

- Pump: The function of this element is to compress the fluid (water vapor) before it enters the boiler. Since the variation in the pressure of the fluid will depend on the speed of rotation of the machine, this pump has been suppressed and only the stresses in the pump piston will be analyzed.
- Intake and regulation system: The speed regulator has a dynamic functionality as it will rotate at a speed proportional to that of the crankshaft and this will create more or less centrifugal force, so it would make more sense to analyze this component within a dynamic analysis. Similarly, the entry of steam through the intake housing, in which the flow control valve is located, will be eliminated from the analysis to simplify the model, since this could be analyzed separately simply by applying a pressure load on the contours of the elements through which the fluid passes.

- Union elements: Many fixing elements have been suppressed and replaced by contact relationships of the 'bonded' type, that is, by welded unions. This type of contact establishes that the nodes of the meshes of the elements whose surfaces are in contact will not have relative movement between them on these surfaces. Given that in most cases, welded joints have a higher resistance than bolted joints, this simplification will make sense. This can be achieved as long as the von Mises stresses (which will be used as the failure criterion by comparing them with the yield strength) in the joint areas are far from failure (safety factor greater than unity). This will ensure that the tension in these zones continues to work in safe conditions and there is no plasticization in the joints. That is to say, if the welded joint resulted in a stress lower than the yield strength, the stress that would occur in the case of a bolted joint would be slightly higher. In Figure 5, the simplified model is shown.

Figure 5. Simplified model of the assembly: front view (**left**) and rear view (**right**).

2.2.2. Assignment of Materials

The assignment of materials to the parts during the 3D CAD modeling with their corresponding mechanical and thermal properties is essential, since it will directly influence the behavior of the element in the analysis to be performed.

If during the modeling of each of these, the material is assigned to them, when opening the Autodesk Inventor Nastran environment, in the operations tree in the 'Model' tab, the idealizations of the materials can be found. For the invention under study, the assigned materials are the same as those assigned to each piece by the author of the plans: steel, carbon steel, stainless steel, cast iron, brass, cast bronze, rubber and nylon. Table 1 presents the values of the properties of the materials used in the analysis.

Table 1. Properties of each material used in the analysis.

Material	Young's Modulus (MPa)	Poisson Coefficient	Density (Kg/m^3)	Yield Strength (MPa)
Carbon Steel	200,000	0.290	7850	350.00
Stainless Steel	193,000	0.300	8000	250.00
Cast Iron	120,500	0.300	7150	758.00
Brass	109,600	0.331	8470	103.40
Cast bronze	109,600	0.335	8870	128.00
Nylon	2930	0.350	1130	82.75

2.2.3. Application of Contacts

For a correct simulation by the finite element method, it must be established what types of contacts there are between the elements that make up the assembly. Autodesk Inventor Nastran allows existing contacts to be established automatically, which is very advantageous since for assemblies with a high number of components, the number of contacts to be established manually would be very high and would take too much time. Therefore, it was decided to automate this process. The existing contacts in the machine are briefly explained below:

- 'Separation' type: This is the most common contact between two elements. In this type of contact, the relative movement between the nodes of the elements in contact is allowed, but with a coefficient of friction (which will be that of the materials). Apart from that, the coefficient of friction depends on other parameters, such as the surface finish of the surfaces in contact, temperature, surface roughness, etc. To simplify, a coefficient of 0.25 was applied between surfaces without lubrication, and a coefficient of 0.1 between surfaces with lubrication, for example, in bearings.
- 'Bonded' type: In this type of contact, relative movement between nodes is not allowed; that is, it behaves as if the elements in contact were welded.
- 'Symmetric/Unsymmetric contact' penetration type: In the case of symmetric penetration, the nodes of a mesh cannot penetrate the nodes of the adjacent mesh, and in the case of the asymmetric type, the penetration of nodes in adjacent meshes in contact is allowed.

One drawback that appears with the generation of contacts automatically is that the software will only make them of one type. Therefore, the behavior of the resulting mathematical model would not correspond to the behavior of the real model, since there are different types of contact in the assembly.

To solve this problem, there are two possible solutions: the first one is to establish the contacts manually, a long and tedious process since there is a large number of components, and the second option is to establish the contacts automatically and modify those that require it. The latter is the option chosen in this investigation, automatically establishing 'bonded'-type contacts, and modifying them to a 'separation'-type contact for those that require it. In this way, a response of the mathematical model that corresponds to that of the real model is achieved.

Figure 6 shows an example of the modification of a bonded contact to a separation contact (a pop-up is displayed to change the contact type between a bearing and the head of a beam).

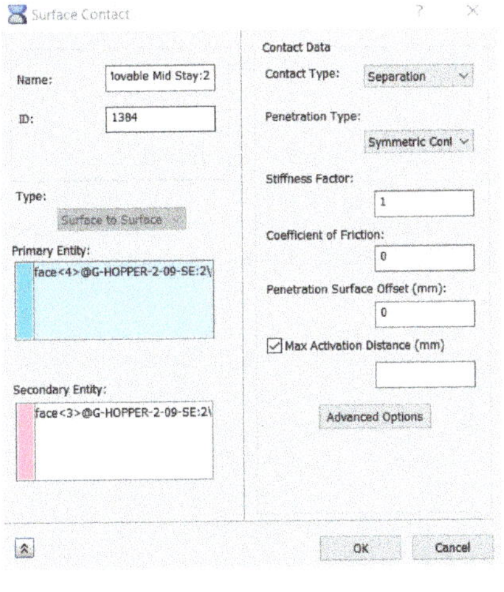

Figure 6. Pop-up to modify the contact type.

2.2.4. Boundary Conditions

In order to carry out a coherent analysis, the boundary conditions that will restrict certain degrees of freedom of the components of the invention must be established, in order to obtain results that are as close as possible to those obtained in real operating conditions.

First, all the lower faces of the bedplate were fixed without any movement (all freedom degrees are restricted) (Figure 7); that is to say, they are prevented from any displacement and rotation. This boundary condition would be equivalent to fixing the bedplate of the steam engine to the ground, or to the chassis of the train or ship where the steam engine was located.

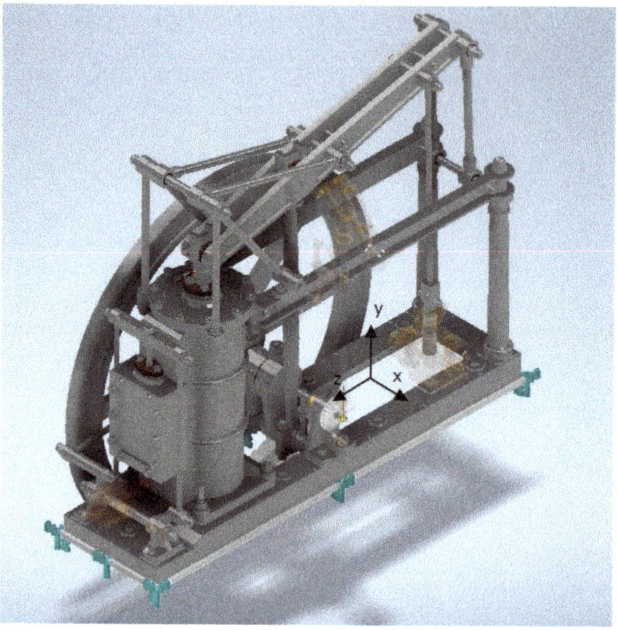

Figure 7. Bedplate boundary conditions.

Secondly, as the pump has been eliminated in the simplified model for analysis, the movement of its piston must be restricted (Figure 8), since it would have an alternative movement in the vertical direction with respect to the pump. Therefore, all displacements are restricted except for the displacement, whose direction coincides with the longitudinal axle of the pump piston. In turn, since two cap nuts that hold the pin that joins the pump piston to the rod have been eliminated in order to simplify the analysis, horizontal displacement (direction of the axis of revolution of the pin) is prevented in order to obtain an operation similar to that which would be obtained in the case of carrying out the analysis with these cap nuts. In Figure 9, the restricted degrees of freedom for the piston pump (those with a tick) are shown.

Finally, since a linear static analysis is going to be carried out simulating the two most critical positions (piston plunger at lower dead center and at upper dead center), the flywheel is fixed (all freedom degrees are restricted) (Figure 10). In this way, the invention will be blocked and the output variables (stress, displacement and safety factor) will be analyzed in both critical positions. In Section 2.2.6, these critical positions are explained in detail, as well as the reason for choosing them among all the possible ones.

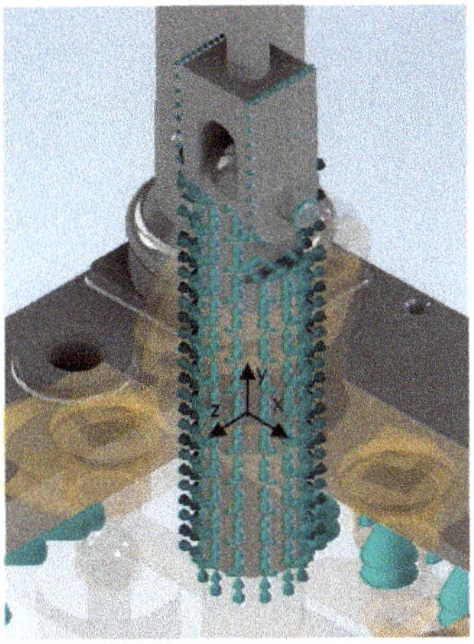

Figure 8. Pump piston boundary conditions.

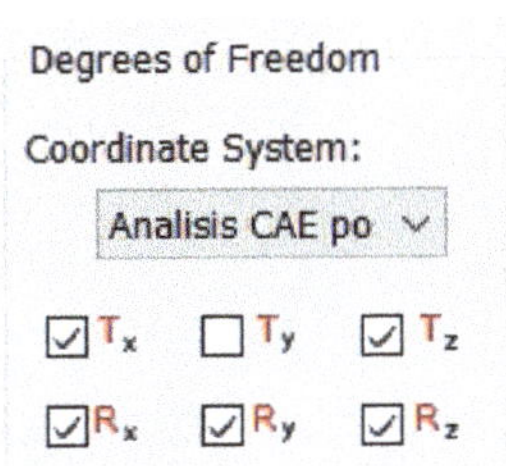

Figure 9. Restricted degrees of freedom for the piston pump.

Figure 10. Flywheel boundary conditions.

2.2.5. Discretization

To finish with the steps prior to the analysis, the continuous medium must be discretized. Discretization consists of dividing the continuum medium into numerous elements defined by nodes. In this way, the analysis variables will be calculated in the nodes of the elements, and the values will be interpolated in them in order to obtain the value inside the element. Depending on the number of nodes that the element has, the interpolation order will vary.

This discretization is necessary in order to be able to carry out the analysis, since the differential equations that govern the behavior of the continuous medium cannot be solved. Therefore, the finite element method is used to obtain an approximate solution to these equations. The error committed between the analysis by the finite element method, and the real case for solving the differential equations, will depend on the grid size.

After this brief description of the finite element method and the need to apply it in order to solve the problem, the discretization for the case study is presented (Figure 11).

Figure 11. Axonometric view of the discretization of the assembly.

In the present investigation, quadratic interpolation order elements were used, that is, parabolic-type elements (tetrahedra) (Figure 12), to obtain more precise results inside the finite element. The software automatically generates a smaller mesh in smaller parts so that the grid fits better to the geometry of the part and the results are more accurate. Thus, the maximum element size (0.029 m) and the maximum element growth ratio (1.1) have been defined. This configuration allows greater control over the mesh of the parts with contacts, meaning less distortion of the elements. In addition, Autodesk Inventor Nastran allows a more advanced configuration of the finite elements, being able to modify the maximum and minimum angles of the triangle of the tetrahedron. Moreover, the growth rate of the element can be modified in contact zones between different solids, as well as in edges or vertices where a smaller element is required for a better adaptation to the geometry of the solid. In this way, a higher-quality mesh is achieved, and, as they are tetrahedrons, a better adaptation to the geometry of each solid. In this investigation, the analysis used a mesh size of 1,132,720 elements and 1,696,754 nodes.

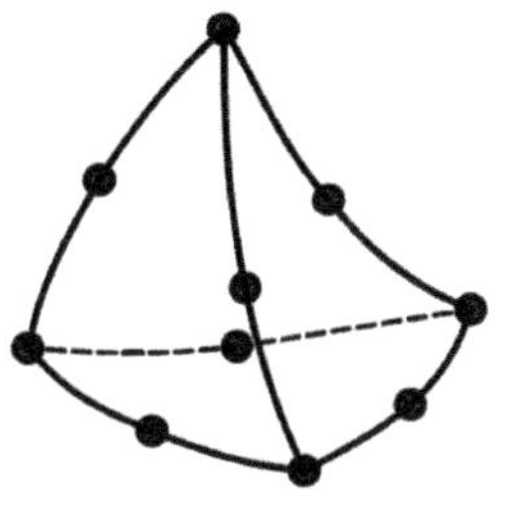

Figure 12. Finite element used. Tetrahedron of quadratic interpolation.

2.2.6. Critical Positions

The piston plunger positions that give rise to the greatest von Mises stress and the largest magnitude displacements should be analyzed, since although many positions could be analyzed, the ones of greatest interest are those that will subject the machine to the conditions of most critical operation.

For the case study, two critical positions are presented (Figure 13), corresponding to the position of the piston plunger at the lower dead center and the upper dead center after closing the steam intake by the slide valve in order to later expand inside the cylinder chamber and produce work. In addition, since it is a double-acting steam engine, both situations must be taken into account. In these positions, the maximum pressure is exerted on one of the faces of the piston and atmospheric pressure on the opposite face, since when the steam is renewed in one of the chambers, the discharge or exhaust occurs in the other, and at that instant, the pressure on the face of the piston and cylinder chamber is atmospheric pressure. Once the piston reaches one of these two critical positions, the pressure inside one of the chambers is at its maximum, and the pressure will produce greater stresses and deformations.

Moreover, since the flywheel is blocked, it will be possible to simulate in this way the placing of the machine in operation and observe how the force is transmitted from the piston to the rest of the elements of the steam engine. This blocking situation can be assimilated to a very common test in reciprocating and rotary internal combustion engines, since the torque present on the engine shaft is measured by braking the flywheel. Figure 13 shows the invention in both critical positions, clearly observing the position of the elements, particularly that of the balance beam, since it is the most innovative element of the invention for the time in which it was developed.

Figure 13. Axonometric views of the critical positions: lower dead center (**left**) and upper dead center (**right**).

2.2.7. Modal Analysis

Every mechanical system has natural frequencies of vibration and has as many modes of vibration (natural frequencies) as degrees of freedom. Each vibration mode is characterized by a frequency (Hz) and a vibration mode (mode number). This is the consequence of subjecting the system to an imbalance caused by the application of an external force. If this external force takes the solid to a deformed position (within the linear elastic regime, without presenting permanent plastic deformations) and suddenly disappears, the solid will return to its original equilibrium position, but it will arrive with certain kinetic energy (since when it stops the force, the potential energy stored by the material is transformed into kinetic energy) that will cause deformation to occur in the opposite direction to that created by the external force, returning to seek the equilibrium position, and thus producing vibration. When the system vibrates, it will do so at a series of frequencies, these being the natural frequencies.

Therefore, since a linear static analysis is to be carried out later, it does not make sense to execute it if any of the vibration modes of the system (steam engine) corresponds to a frequency of 0 Hz, since it would behave like a mechanism. In this way, for both critical positions of the assembly, the first 10 natural frequencies were obtained, and it was verified that all of them are greater than 0.

2.2.8. Linear Static Analysis

As previously mentioned, the analysis to be carried out will be linear static, after verifying that the system does not have any vibration mode corresponding to the frequency of 0 Hz. This analysis will force a convergence analysis of the mesh to ensure that the results obtained are correct and that this convergence approaches the real solution. Therefore, the objective of the convergence analysis is to observe for what value of the element size the von Mises stress does not vary to a large extent.

Determination of the Strain Envelope

The correct determination of the strain envelope is essential for a correct analysis that does not lead to erroneous results.

The plans [7] that have made it possible to obtain the 3D CAD model of the invention show its reduced dimensions. As mentioned previously, the main application of this machine was its use in small boats and trains, so its dimensions would be larger depending on the size of the means of locomotion.

Despite all this, and as a starting point, an analysis was carried out for a manometric working pressure of 0.75 MPa, since exact information on the operating conditions of the machine is not available. The steam conditions at the cylinder inlet depend on the geometry of the machine, that is, the dimensions of the plunger, the chamber, etc., as well as the application. For use in maritime vessels or trains, the pressure value could range between 1 and 8 MPa, depending on the machine, and in power plants for the generation of electrical energy, the outlet pressure of the boiler could have a value greater than 10 MPa. This pressure is produced by the steam that comes out of the boiler, being applied in the slide valve chest block, the slide valve and the corresponding piston face, depending on the position. Therefore, this manometric pressure value is used, because throughout the invention in its environment, and in the chamber opposite to the pressure application chamber, there will be an applied pressure of a value equal to atmospheric pressure. In addition, after expanding the steam inside the chamber, the exhaust valve opens and the steam escapes at a constant pressure equal to atmospheric pressure, while in the other chamber, the admission of steam will be taking place at the working pressure.

After the application of the pressure of 0.75 MPa, the behavior of the machine can be observed, and if the minimum safety factor is less than the unity, the steam engine will have exceeded the yield strength and therefore, it will fail at the first cycle static charge. In this way, several analyses must be carried out until a pressure value is determined that causes a maximum von Mises stress lower than the yield strength value, and whose safety factor is greater than the unity for the machine to work safely (values between two and four are those used when designing machines today).

In Figure 14, the application of the pressure loads for the critical position corresponding to the position of the piston plunger at the lower dead center can be seen in detail.

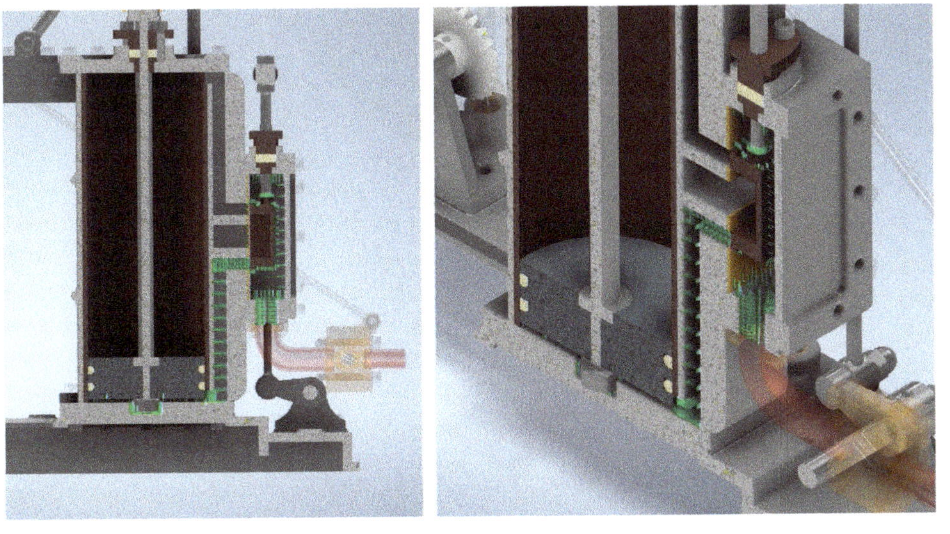

Figure 14. Pressure loads at the lower dead center position: front view (**left**) and axonometric view (**right**).

In the same way, the pressure load for the critical position corresponding to the upper dead center would be applied. In that case, the load would be applied to the opposite face of the piston with respect to the lower dead center position. In this way, it will be possible to observe two critical and opposite behaviors, given that for the lower dead center position, the piston rod will work in compression while for the upper dead center, the piston rod will work in traction. Viewed in this way, a more critical behavior can be anticipated in the upstroke of the piston, that is, for movement from the lower dead center to the upper dead center. The fact that the piston rod works in compression instead of traction produces greater criticality in the deformed state, since a compression load can give rise to eccentricity of the applied load if the piston rod deforms, producing buckling. For this reason, it is foreseen that the most critical position will be the position of the lower dead center.

Analysis Execution

After applying the pressure loads, the analysis will be carried out. Autodesk Inventor Nastran allows users to automatically perform several iterations in the simulation to determine the convergence of the mesh. Finally, a graph will be obtained for the convergence analysis as well as the response of the steam engine to the load, quantified by the values of the von Mises stress and by the displacements, as well as the safety factor defined as the quotient between the yield strength of the material and the von Mises stress at a given point.

The steam engine is an assembly with a high number of components, obtaining a very complex mesh with a high computational cost in the simulation. However, the mesh convergence analysis performed by the software itself reaches convergence in two iterations. For this reason, in order to ensure that the results offered by the software are completely reliable, it was decided to carry out a mesh convergence analysis manually, the results being those corresponding to the manual mesh convergence analysis. This analysis consists of performing a first mesh with a determined global mesh size and proceeding to the simulation. Once the results are obtained, the place where the external strains give rise to a maximum von Mises stress can be determined, noting this value together with the iteration number and the element size.

The next step is to perform a local mesh control in the area of maximum stress, reducing the size of the finite element in that area. Control can be achieved in several ways: by refining vertices, edges, surfaces or solids. For the case study, it was decided to completely refine the solid (piston rod) where the maximum stresses are produced. Once the value of the size of the element has been reduced, the meshing of the assembly is carried out again and the simulation is repeated. This process is repeated until the relative error of the von Mises stresses between iteration (i) and iteration (i-1) is less than an established percentage (in the present investigation, a value of 10% was adopted).

3. Results and Discussion

This section presents the results obtained for the modal analysis and linear static analysis in each of the two critical positions studied. It should be noted that the figures presenting the analysis results are screenshots taken from the software (Autodesk Inventor Nastran), which uses the comma as the decimal separator instead of the dot.

3.1. *Critical Position: Lower Dead Center*

3.1.1. Modal Analysis

After applying the corresponding boundary conditions, the simulation is performed in Autodesk Inventor Nastran. Figure 15 shows a graph in which the abscissa axis corresponds to the mode number of the vibration and the ordinate axis to the natural frequency (Hz) corresponding to this mode number.

The graph shows how vibration mode 1 corresponds to a frequency value greater than 200 Hz, this being the lowest of its natural vibration frequencies. Therefore, a static analysis can be performed since it will not behave as a mechanism under the simulation conditions at the critical position corresponding to the position of the piston plunger at the lower dead center.

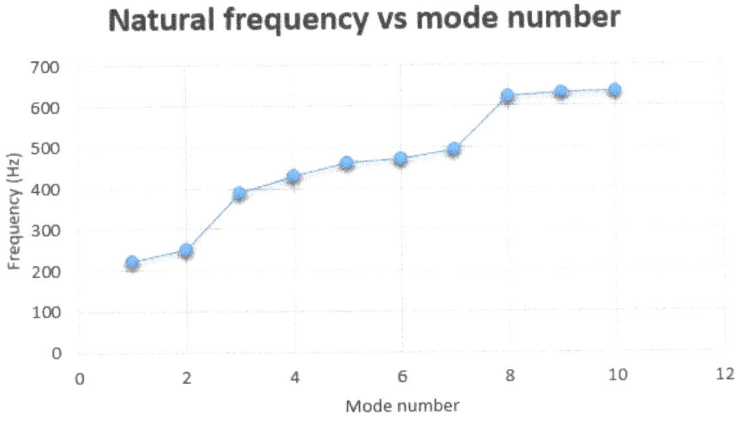

Figure 15. Natural frequency vs. mode number of the vibration.

Thus, the first 10 natural frequencies of vibration for the steam engine at this critical position are 221.17 Hz, 249.73 Hz, 390.74 Hz, 431.69 Hz, 460.54 Hz, 470.16 Hz, 493.69 Hz, 623.09 Hz, 632.22 Hz and 636.54 Hz. A system has as many natural frequencies as degrees of freedom, but it does not make sense to obtain all the natural frequencies since only those with the lowest value are of interest in order to rule out the possibility that the invention behaves like a mechanism (this is for a natural frequency of vibration of 0 Hz) and to carry out static analysis.

3.1.2. Linear Static Analysis

Below, the results are presented for the critical position corresponding to the lower dead center, after closing the steam intake to the cylinder for its subsequent expansion inside the chamber.

As previously explained, first, a linear static analysis was carried out under a pressure of 0.75 MPa, producing a maximum von Mises stress of 1089 MPa at the top of the piston rod (Figure 16).

This value of the von Mises stress results in a safety factor less than the unity (0.51). Therefore, it can be stated that the machine would fail if the steam pressure at the intake was 0.75 MPa. This makes it necessary to carry out another analysis to obtain a lower value of the von Mises stress, particularly 0.15 MPa, and to analyze the results to determine whether this value of steam pressure is adequate for the correct operation of the machine.

The analysis carried out above not only indicates that the machine fails, but also offers information on the place where the maximum von Mises stress is located, particularly at the top of the piston rod. This will make it possible to perform a mesh refinement in that area and to study the convergence of the mesh, comparing the element size with the value obtained from the von Mises stress.

After performing the analysis with different mesh sizes in the piston rod, the convergence analysis graph of the mesh is obtained (Figure 17), as well as the table with the size of the element, the von stress Mises, the relative error (%) and the iteration number (Table 2).

From Table 2, it can be seen that the relative error obtained for an element size of 0.4 mm is 7.35%. The computational limitations for the development of the analysis and the complexity of the assembly do not allow further refinement of the mesh in the piston rod, which is the most critical element. Despite this, an error slightly higher than 7% offers good results, and since the purpose of the analysis is only to obtain information about the design of the invention, this relative error can be accepted.

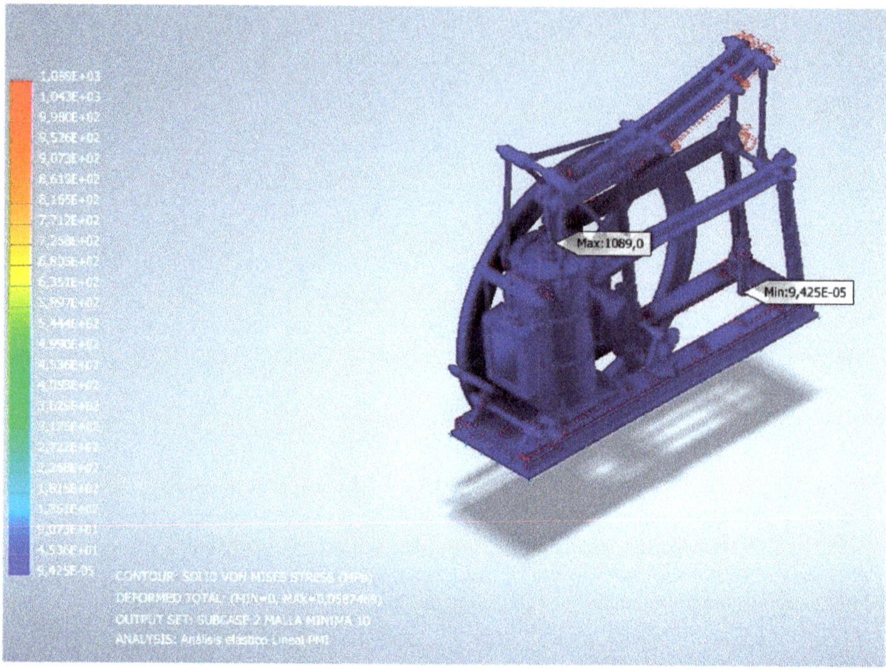

Figure 16. Von Mises stress distribution for a pressure of 0.75 MPa.

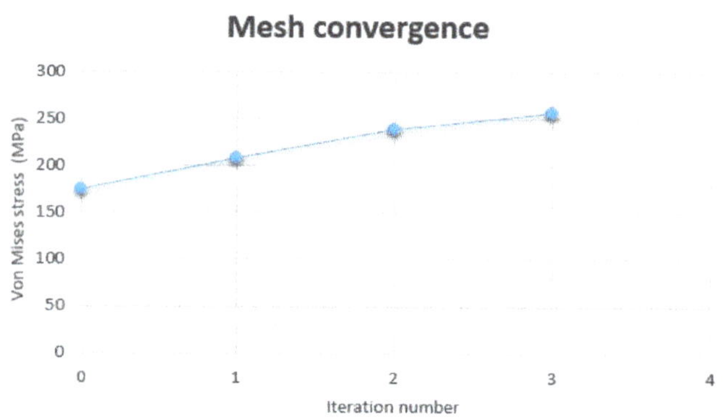

Figure 17. Mesh convergence graph.

Table 2. Mesh convergence analysis.

Element Size (mm)	Von Mises Stress (MPa)	Relative Error (%)	Iteration
1.5	175.5		0
1	208.1	18.58	1
0.5	239.4	15.04	2
0.4	257	7.35	3

Below are shown the von Mises stress distribution (Figure 18) and the exact location (piston rod) of its highest value (257 MPa) (Figure 19). In addition, it should be noted that the von Mises stress, displacements and safety factor distributions were obtained on the deformed position of the invention multiplied by a factor of value 10, in order to clearly observe the shape that the assembly adopts in the deformed position, since for a value factor of the unity, the deformed position could hardly be seen due to the very small value of the displacements obtained.

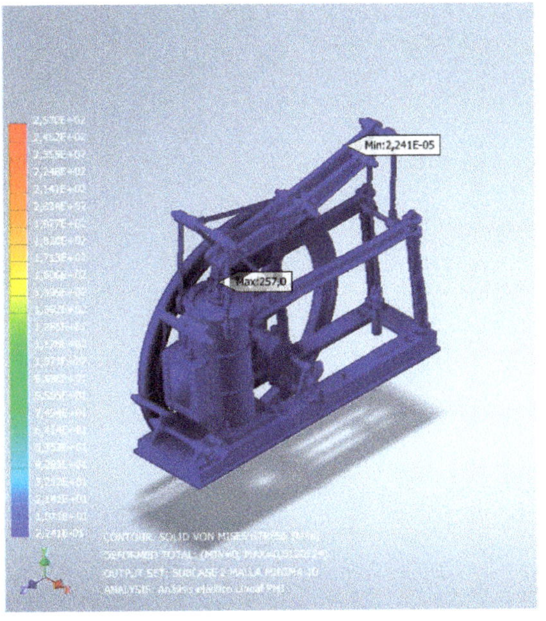

Figure 18. Von Mises stress distribution.

Figure 19. Location of the highest von Mises stress.

On the other hand, Figure 20 represents the distribution of displacements, in which it can be seen that these are negligible.

Figure 20. Displacement distribution.

Figure 21 shows the safety factor, defined as the result of dividing the yield strength of the material by the value of the von Mises stress at a specific point. In this figure, it can be seen that the smallest value of the safety factor is 2.307, located at the piston rod.

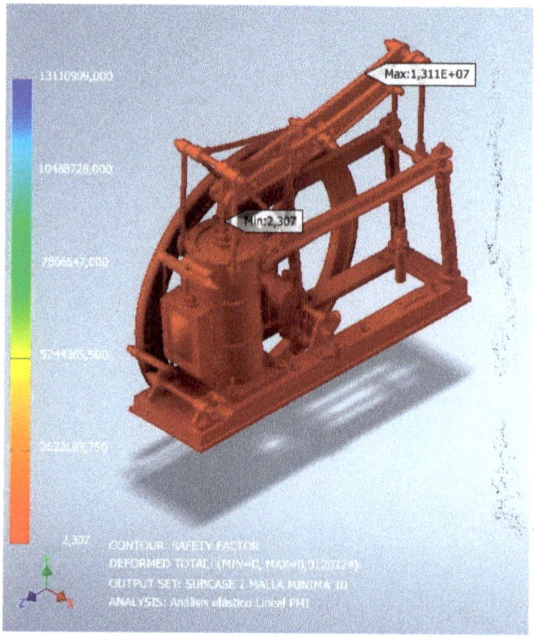

Figure 21. Location of the lowest value of the safety factor.

3.2. Critical Position: Upper Dead Center

3.2.1. Modal Analysis

After applying the corresponding boundary conditions, the simulation was performed in Autodesk Inventor Nastran. Figure 22 shows a graph in which the abscissa axis corresponds to the mode number of the vibration, and the ordinate axis to the natural frequency (Hz) corresponding to this mode number.

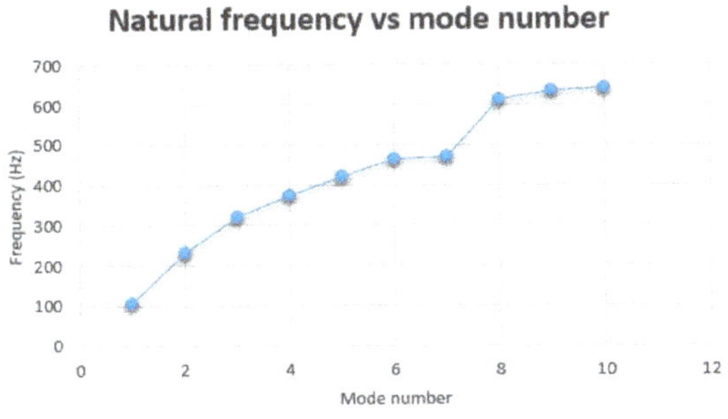

Figure 22. Natural frequency vs. mode number of the vibration.

The values of the first 10 natural frequencies are as follows: 105.34 Hz, 232.40 Hz, 319.77 Hz, 375.03 Hz, 422.13 Hz, 464.58 Hz, 472.74 Hz, 613.48 Hz, 634.45 Hz and 640.17 Hz.

As in the case of the critical position corresponding to the lower dead center, since the natural frequency associated with the first mode of vibration is greater than 0 Hz, linear static analysis can be performed at the critical position of the upper dead center, since this would not behave like a mechanism.

3.2.2. Linear Static Analysis

Next, as for the lower dead center critical position, the value obtained for each of the variables of interest is shown, that is, the von Mises stress, displacement and safety factor, as well as a graph of the convergence analysis of the mesh (Figure 23) and the table with the element size, the von Mises stress, the relative error (%) and the iteration number (Table 3). All these values were obtained for the same working pressure adopted for the lower dead center of value 0.15MPa, it being necessary to verify that, for this critical position, and at said working pressure, the safety factor is greater than the unity, thus ensuring the operation of the steam engine under safe conditions and that there is no failure under the static load of the first cycle.

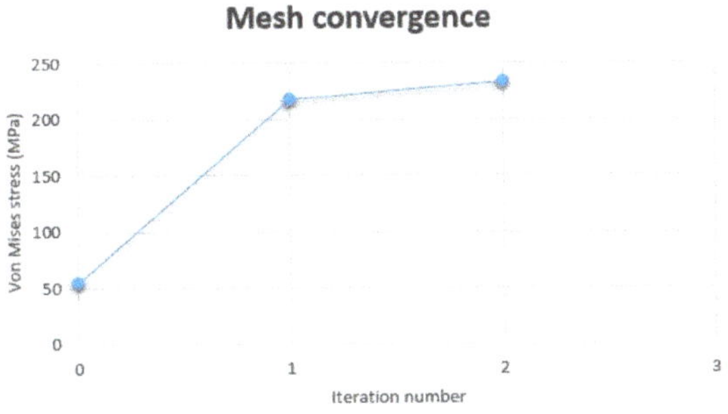

Figure 23. Mesh convergence graph.

These results are the ones obtained for the indicated mesh element sizes, refining the piston rod, since the maximum von Mises stress in the piston rod is obtained in the first iteration (iteration 0). Thus, in the following iterations, the analysis is executed by applying a local mesh control on that piston rod, thereby reducing the size of the finite element until the solution converges (relative error less than 10%).

Table 3. Mesh convergence analysis.

Element Size (mm)	Von Mises Stress (MPa)	Relative Error (%)	Iteration
2	54.9		0
1.5	217.8	296.64	1
0.75	233.4	7.16	2

In the same way as in the lower dead center critical position, and given the computational limitations and the complexity of the machine, the mesh cannot be further refined.

The von Mises stress distribution is shown below (Figure 24), as well as an enlarged view of the location (piston rod) of the highest value obtained (233.4 MPa) (Figure 25).

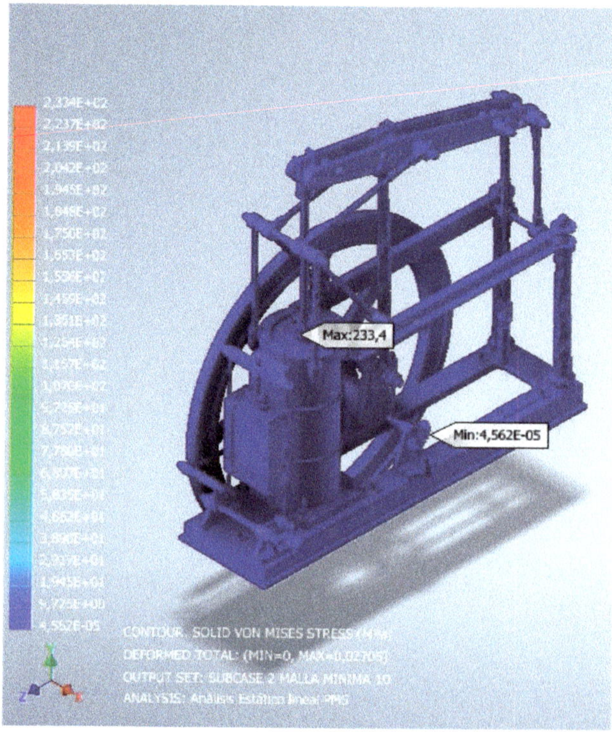

Figure 24. Von Mises stress distribution.

Figure 25. Location of the highest von Mises stress.

Figure 26 represents the distribution of displacements, in which it can be seen that they are negligible.

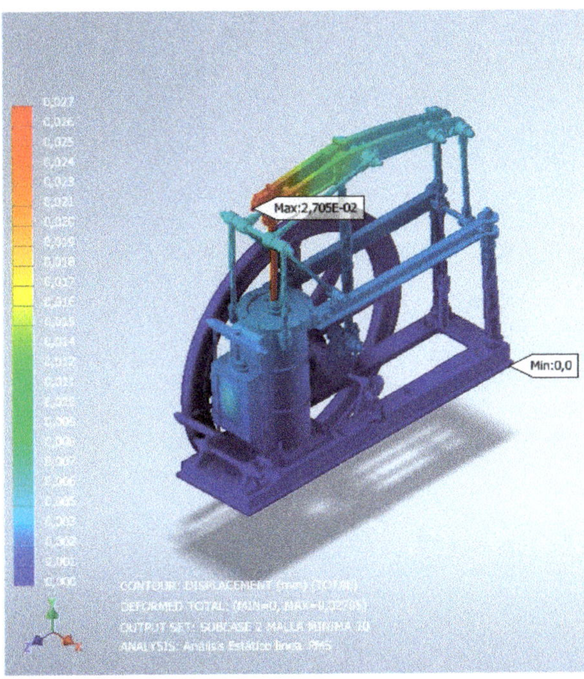

Figure 26. Displacement distribution.

Finally, Figure 27 shows the distribution of the safety coefficient for this critical position and the location of the lowest value of this coefficient (1.445), located at the piston rod.

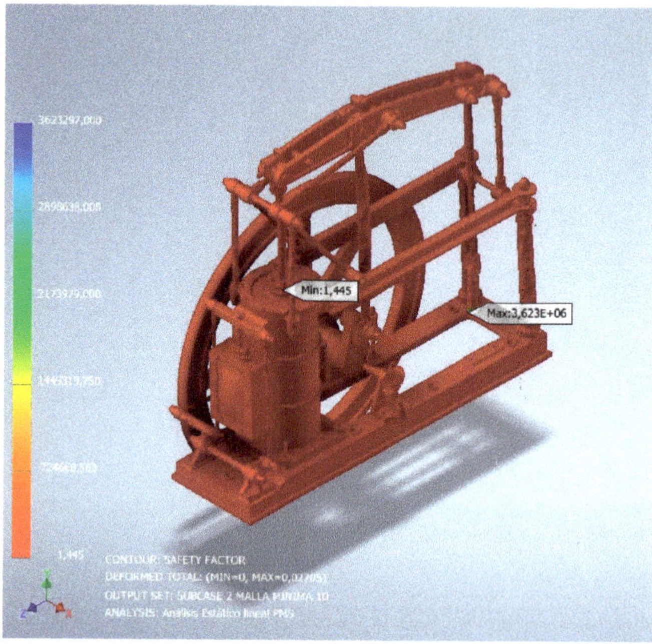

Figure 27. Location of the lowest value of the safety factor.

3.3. Discussion of the Results

According to the results, the highest von Mises stress was obtained in the critical position of the lower dead center. In this position, there is a compression load that can cause buckling of the piston rod. Therefore, in future research, this rod could be analyzed in order to determine whether the rod complies under the buckling failure criterion.

Regarding the displacements, and as previously indicated, an increase factor of 10 was applied in order to more clearly observe the deformed position of the steam engine in both critical positions (lower and upper dead center), so that the representation is not the real one. In both analyses, practically negligible displacements were obtained. This favors the proper functioning of the steam engine under this pressure load, since the hypothesis of small deformations can be applied and the strains in the undeformed position can be calculated (that is, using the Eulerian formulation method for meshing, which reduces the computational cost). In addition, since there are no large deformations, the relative movement between the elements will give rise to the least possible friction, since, for example, if there were large deflections in the piston rod, there would be greater friction contact in the bushing and piston seal itself, and mechanical energy would be lost.

Finally, as desired, the safety factor is greater than the unity, so it can be stated that the machine does not fail under the first cycle static load at a working pressure of 0.15 MPa. However, even though the maximum von Mises stress at the upper dead center is less than that obtained at the lower dead center, the safety factor is also less. This is due to the fact that higher stress is produced in a secondary element than in the lower dead center analysis, and its mechanical characteristics (Young's modulus and yield strength) are lower than those of the piston rod.

Therefore, with the present research, it has been possible to determine the maximum admissible steam pressure in the admission (working pressure) for this machine, corresponding to 0.15 MPa. This value can be used to determine the working pressure in other similar machines, relating the parameters of both models through dimensional analysis.

4. Conclusions

In this paper, the design of the single-cylinder steam engine of the Grasshopper beam by Henry Muncaster according to material resistance criteria was analyzed. In the absence of detailed information on the operating data of the invention, this double-acting steam engine was studied with computer-aided engineering techniques, specifically through linear static analysis, in the two critical positions: when the piston rod is at the lower dead center and at the upper dead center. This was possible thanks to the Autodesk Inventor Nastran software.

The objective of this research was to study from the design point of view what would be the maximum admissible steam pressure in the admission (working pressure) so that the invention would not fail, according to material resistance criteria, thus analyzing von Mises stresses, displacements and safety factors.

Thus, after manual refining, a mesh convergence analysis was performed, obtaining a relative error of 7.35% and an element size of 0.4 mm for the position of the piston plunger at the lower dead center, and a relative error of 7.16% and an element size of 0.75 mm for the upper dead center position. The iterative process then stopped at the fourth iteration for the lower dead center position, and at the third iteration for the upper dead center position, taking into account that the first iteration was carried out to estimate the zone of greatest criticality of the invention in both positions.

On the other hand, the highest von Mises stress for the working pressure (0.15 MPa) obtained a value of 257 and 233.4 MPa in both critical positions (lower and upper dead center, respectively), being located on the piston rod. Furthermore, the minimum safety factors were greater than the unity (2.307 and 1.445 in lower and upper dead center, respectively), guaranteeing that the invention did not fail.

In addition, and as possible future research, a dimensional analysis could be carried out from the determination of the working pressure (0.15 MPa) that relates this pressure to the relevant dimensions of the machine, such as the diameter and stroke of the piston plunger, in order to be able to extrapolate results to a similar invention (same geometric characteristics but different dimensions). This would allow the number of analyses to be reduced, since carrying out tests for a single machine would allow values to be extrapolated to other similar machines. Another advantage of applying the dimensional analysis technique to this simulation is that it not only determines the maximum operating pressure in a similar machine, but also that an inverse study can be carried out if it is required for a certain function, such as calculating the specific value of torque on the flywheel of a train that circulates at a certain speed. In this way, and based on these data, the working pressure that produces the torque is determined, analyzing the thermodynamic cycle and applying the most appropriate and effective mechanical and thermal performance and then determining the geometry necessary to produce this torque and/or effective power.

Author Contributions: Conceptualization, J.I.R.-S.; methodology, J.I.R.-S. and J.F.G.-A.; investigation, J.I.R.-S. and J.F.G.-A.; formal analysis, J.I.R.-S. and J.F.G.-A.; visualization, J.I.R.-S. and J.F.G.-A.; supervision, J.I.R.-S.; writing—original draft preparation, J.I.R.-S. and J.F.G.-A.; writing— review and editing, J.I.R.-S. and J.F.G.-A. All authors have read and agreed to the published version of the manuscript.

Funding: The research presented in this paper was possible thanks to a collaboration grant with the Department of Engineering Graphics, Design and Projects of the University of Jaen obtained in the 2022 call from the Ministry of Education and Vocational Training of the Government of Spain.

Data Availability Statement: Not applicable.

Acknowledgments: We would like to thank the anonymous reviewers of this paper for their constructive suggestions and comments.

Conflicts of Interest: The authors declare no conflict of interest.

References

1. Inkster, I. (Ed.) *History of Technology*; Bloomsbury Academic: London, UK, 2004; Volume 25.
2. Jenkins, R. *Links in the History of Engineering and Technology from Tudor Times*; The Newcomen Society at the Cambridge University Press: Cambridge, UK, 1971.
3. Russell, B. *James Watt: Making the World Anew*; Reaktion Books: London, UK, 2014.
4. Who Was Henry Muncaster? Available online: https://modelengineeringwebsite.com/Henry_Muncaster.html (accessed on 29 June 2023).
5. Muncaster, H. *Model Stationary Engines—Their Design and Construction*; TEE Publishing Ltd.: Oxford, UK, 1912.
6. Grasshopper Beam Engine. Available online: https://en.wikipedia.org/wiki/Grasshopper_beam_engine (accessed on 29 June 2023).
7. Single Cylinder Grasshopper Beam Engine. Available online: https://modelengineeringwebsite.com/Muncaster_grasshopper.html (accessed on 29 June 2023).
8. Rojas-Sola, J.I.; Gutiérrez-Antúnez, J.F. Geometric modeling and digital restitution of a single cylinder steam engine of the Grasshopper beam by Henry Muncaster. In *Advances in Engineering Design IV*, 1st ed.; Manchado del Val, C., Suffo Pino, M., Miralbés Buil, R., Moreno Sánchez, D., Moreno Nieto, F.D., Eds.; Springer: Cham, Switzerland, 2024; (*accepted for publication*).
9. Crowley, T.E. *The Beam Engine: A Massive Chapter in the History of the Steam*, 1st ed.; Senecio Publishing: Oxford, UK, 1982; pp. 64–65.
10. Semmens, P.W.B.; Goldfinch, A.J. *How Steam Locomotives Really Work*; Oxford University Press: Oxford, UK, 2003; p. 97.
11. Liu, Y.; Huang, M.; An, Q.; Bai, L.; Shang, D.Y. Dynamic characteristic analysis and structural optimization design of the large mining headrame. *Machines* **2022**, *10*, 510. [CrossRef]
12. Stavroulakis, G.E.; Charalambidi, B.G.; Koutsianitis, P. Review of computational mechanics, optimization, and machine learning tools for digital twins applied to infrastructures. *Appl. Sci.* **2022**, *12*, 11997. [CrossRef]
13. Jumper, L.; Shih, R.H. *Parametric Modeling with Autodesk Inventor 2023*; SDC Publications: Mission, KS, USA, 2022.
14. Autodesk Inventor Nastran 2023. About Tutorials. Available online: https://help.autodesk.com/view/NINCAD/2023/ENU/?guid=GUID-DB7160BE-0C72-47B9-B5EF-FC4925B455CE (accessed on 11 June 2023).

machines

MDPI

Article

Green Machining of NFRP Material

Zuzana Mitaľová [1,*], Dagmar Klichová [2], František Botko [1], Juliána Litecká [3], Radoslav Vandžura [1] and Dušan Mitaľ [1]

1 Faculty of Manufacturing Technologies, Technical University of Košice, 080 01 Prešov, Slovakia; frantisek.botko@tuke.sk (F.B.); radoslav.vandzura@tuke.sk (R.V.); dusan.mital@tuke.sk (D.M.)
2 The Czech Academy of Sciences, Institute of Geonics, 708 00 Ostrava, Czech Republic; dagmar.klichova@ugn.cas.cz
3 Faculty of Humanities and Natural Science, University of Prešov, 080 01 Prešov, Slovakia; juliana.litecka@unipo.sk
* Correspondence: zuzana.mitalova@tuke.sk; Tel.: +421-55-602-6465

Abstract: Nowadays, great emphasis is placed on environmental aspects of production processes with focus to lower carbon footprint. Natural fibre-reinforced plastics (NFRP) show potential for application in many fields of industry due their specific properties. Machining of NFRP-based materials is meeting several problems arising from non-homogenous structure as well as plastic-based matrix. Machining of NFRP using conventional technologies meets limitations due to the properties and geometry of the tools. Abrasive water jet (AWJ) machining can solve some of the problems machining NFRP materials. The presented article focused on surface topography evaluation of one kind of NFRP composite material after cutting by AWJ. Optical profilometry and 3D microscopy were applied for measurement of surface roughness parameters of surfaces created by AWJ with variable cutting parameters. Maximal height of profile *Rz* was measured in 20 lines perpendicular to the jet direction form upper to lower cut line. Structure of cut surface was observed and evaluated for different technologic parameters. The obtained results show promising presuppositions for application of AWJ technology for cutting of NFRP based materials.

Keywords: water jet cutting; non-conventional machining; NFRP composite; natural fibre-reinforced plastics; wood plastic composite; WPC

1. Introduction

The current interest in the use of natural materials as a substitute for fossil raw materials is justified from a material, technological, financial and, last but not least, environmental point of view (reduction in carbon footprint—CO_2). The trend was long supported by the European Commission (the *European Green Deal* programme for minimising the carbon footprint and reducing emissions in the long term). In the last decade, NFRP (natural fibre-reinforced polymer) materials replaced several composite materials reinforced with synthetic fibres, the recycling of which is difficult (e.g., glass fibres—for comparison of the characteristics of synthetic versus natural fibres—see Table 1). NFRP composite materials contain natural fibres and a polymer matrix in a specified ratio (+ additional substances—additives). Thermoplastics materials usually dominate in matrices for NFRP: polypropylene (PP), polyethylene (PE), polyvinylchloride (PVC), polyamide (PA). On the other hand, thermoset materials like phenolic, epoxy and polyester are the most commonly used matrices [1].

The lignocellulosic fibres in the polymer matrix are obtained from cotton, loofah, sisal, flax, hemp, ramie, coconut, bamboo, banana, pineapple, palm leaves, rice, maize, barley husks and deciduous/coniferous trees (Figure 1 shows the chemical composition of the lignocellulosic fibres—consist of lignin, hemicellulose, cellulose). Their advantages include non-toxicity, relatively high strength, low weight/low price independent of the oil price and biodegradability. The resulting NFRP properties of the composite material depend on the percentage (%) in which the of individual components account for the total volume, their interfacial adhesion, production technology, fibre origin or production/finishing technologies.

Table 1. Comparison of selected characteristics of synthetic and natural fibres [2].

Characteristics	Natural Fibres (NF)	Synthetic Fibres (e.g., Glass Fibre)
Cost	↓ Low	↑ higher compared to NF
Density	↓ Low	↑ double compared NF
Recyclability	✓ Yes	✘ No
Renewability	✓ Yes	✘ No
CO_2 neutrality	✓ Yes	✘ No
Abrasion to machines	✘ No	✓ Yes
Health risk associated with production	✘ No	✓ Yes
Disposal	Biodegradable	Non-biodegradable

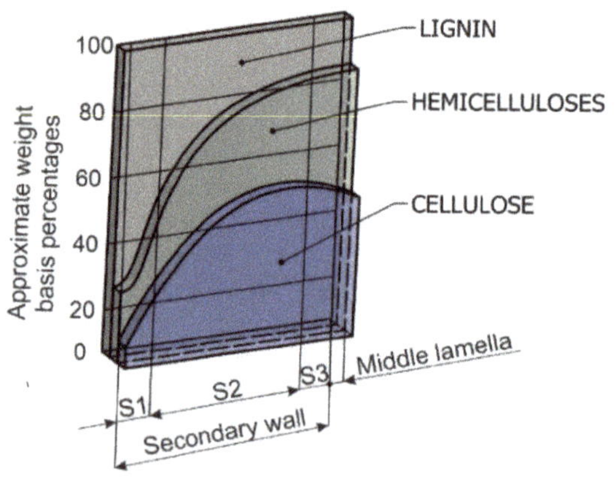

Figure 1. Chemical composition of a typical lignocellulosic fibre (lignin + hemicellulose + cellulose in percentual ratio) [3].

The first NFRP composite material for manufacture of vast numbers of boards/tubes for electronic purpose was used as early as in 1908 (sheet from phenol/formaldehyde resins reinforced with paper or cotton). Due to a reduction in vehicle weight, CO_2 emissions and price, NFRP composite materials were also applied in the automotive sector. In 2012, more than 124,000 tonnes of natural fibre composites were shipped globally for automotive purposes. The available research shows that the use of selected types of natural fibres increased the mechanical properties of the resulting parts (jute, hemp, flax, banana and bamboo fibres—the mechanical properties listed in Table 2) [4,5].

Table 2. Mechanical properties of selected natural fibres [6–9].

Fibre Type [1]	Tensile Strength [MPa]	Tensibility [%]	Young's Modulus [GPa]	Study
Jute fibre	393–773	1.5–1.8	26.5	Kumar and Sharma, 2007 [6]
Hemp fibre	550–900	2.0–3.0	70	Lu et al., 2012 [7]
Flax fibre	800–1500	2.7–3.2	60–80	Shashria, 2019 [8]
Banana fibre	540–900	-	34.8	Narayanan and Elazaperumal, 2012 [9]

[1] Values may vary depending on the type and origin of the fibres.

The composite retains its toughness and is suitable for visually as well as structurally functional parts—e.g., door panels, columns, consoles, etc. Adding wood fibres (particles in the form of flour) to PE/PP/PVC yields a material applied in the construction industry—wood plastic composite (in short, WPC).

As a filler, the composite is typically used in wood plastic composite product, a wood flour obtained from maple, pine, cedar, spruce-fir or oak (determined by geographical location, type, price and availability). The composite product subjected to extreme moisture can be applied to obtain a flour extract from red maple—which improves water resistance of used composite material. Or, to improve durability, it is possible to use a resin of Guayele plants: *Parthenium argentatum* (normally obtained as an additional product when processing rubber). Hardwood provides better tensile properties and heat deflection when compared with softwood. The wood flour presents the crushed wood grains and there is appearance similar to a standard flour—this is a term commonly used in the practice. There are two steps in wood production: first step—size reduction (using a hammer mill/attrition mill or chipper) and second step—a size classification (screening by sieves). In the process of wood flour production, it is necessary to include the drying phase. The usual content of water in wood fibre/flour is about 5–15%, where increasing humidity reduces mechanical properties and thermal stability of the final composite product. For wood flour, it is necessary to keep the moisture content below 1% (in relation to this fact, it is requisite to storage flour in closed plastic bags, until mixing with polymer). The particles are sorted by a size—using sieves with different mesh—and are then classified according to US standard (for example: a particle with a diameter of 850 μm with an identification of 20 MESH/a particle with a diameter of 250 μm with an identification of 60 MESH according to the standard). In the practice, flour with a size between 60 and 80 MESH can be used. For WPC production, it is possible to apply waste from the woodworking industry, too. For example: a co-product like trimming for sawmills, breakdown of urban and demolition wood, or logging trimmings/slash. Adding starch particle into a matrix helps to prevent the degradation (starches from potatoes, rice, grain from cereals). During the melting process, it is necessary to ensure that the temperature does not exceed 200 °C, though some studies state 180 °C, because it would cause wood decomposition. Due to the thermal stability of wood, thermoplastics are used because they can be processed at relatively low temperatures below wood's thermal degradation. Thermoplastics are polymers, which turn to a liquid when heated and a freezy-to-glassy state when cooled. High density polyethylene (HDPE), polypropylene (PP) and polyvinylchloride (PVC) are the most common thermoplastic polymers used in WPCs (according to Klyosov, 2007). Currently studies are pointing to the application of polymethyl methacrylate (PMMA) or Nylon 6. Similarly, biopolymers and biodegradable polymers can be applied as an adequate substitute for thermoplastics (PLA—polylactic acid, or based on polyhydroxyalkanoates—PHB/PHBV). PLA and PHB matrices have similar properties to polypropylene matrices. Additives improve mechanical properties of a final product, provide a chemical stability and more easily process (coupling agents, lubricants, stabilizers, flame retardants, biocides, pigments, fillers, chemical and physical blowing agents) [1,10–14].

The final properties of WPC products depend on several attributes:

- Type of applied polymer matrix;
- The percentage of organic reinforcement (type of wood or plant fibres), the morphology of particles, their physical properties and moisture content;
- Percentage of individual additives;
- Technology and conditions of production process;
- The origin of raw components (possibility of applied plastic recycled, materials, geographical location, etc.);
- Interaction between components.

General WPC production processes are [1,10]:

- Extrusion—for linear profile (technology applied in up to 97% of manufactured WPC products);

- Injection moulding—for 3D parts of regular or irregular shapes, application in mass production, suitable for polymers with low molecular weight. The production process is fundamentally like the production of injected plastics, (disadvantages of the technology: high procurement costs, imperfect compounding of the components of WPC composite material/advantages: minimal waste, short cycle time, observe required dimensional tolerances);
- Compression moulding—cost-effective production of complex parts (disadvantages of technology: complex mould design, the ratio of applied resin and fibre is difficult to control, advantages: no need of qualified staff);
- Calandering—special way of "rolling" in floor production.

Applying thermoplastic matrix offering usage of FDM technology (fused deposition modelling), using direct layering of melted material via heated 3D print head.

WPC composites are applied to flooring, outdoor bed boards, park benches, building templates, production of pallets, fencing and the automotive industry (especially in China). European and American companies recently saw an increase in consumer demand for these building materials. The European WPC market was estimated at over 450 kilotons in 2021 [1,10,15].

The European market is dominated by Tecnaro GMBH, Jelu-werk J. Ehler gmbH & Co. KG, Novo-Tech Trading GmbH & Co. KG. Currently, NFRP composites are used for packing, too. For example, ENKEV produced a product such as Cocolok made from natural fibres/coconut and latex, or applied starch-based composite for tableware. NFRP composite materials are also used in the production of canoes, sporting goods, musical instruments and the transport industry, too [16,17].

Conventional NFRP machining is a long-lasting problem. The existing literature demonstrated that the machinability of composite materials is not the same the machining of homogeneous materials (compared to conventional metal working) [18].

The reasons for this are: molecular nature of the composite material, its inhomogeneity (heterogeneity), fibres pull out during drilling operations, low production productivity, poor technological inheritance indicators values—for example, surface roughness. The resulting characteristics (*Rz*—maximum height and *Ra*—average roughness) depend on the variable conditions of the production technology and the geometry of the applied tool [19–23].

The possibility of resolving the problem is by machining by the AWJ technology. The technology itself offers the following advantages: absence of heat-affected zone (HAZ), no potential fire hazard, low stress on the workpiece, higher flexibility and productivity and no aerosol generation; hence, AWJ machining is called green machining. Machining of NFRP composite materials using the AWJ technology depends on several process variables such as hydraulic (water) pressure, type of the NFRP material, nozzle distance, abrasive type/abrasive size, mass flow rate, etc. The output parameters (surface roughness, material removal rate, kerf taper angle) of the AWJ cutting process are related to the input parameters [24–27].

Researchers examined machining of "green composites" by AWJ/WJ, but very few of them addressed machining of the NFRP composite materials filled with wood. One of the latest studies is the work of Boopathi et al. (2022), dealing with the impact of the input parameters traverse speed and water jet pressure on the output characteristics—surface roughness SR and kerf angle KA (using the Taguchi method), when machining a composite material with neem wood saw powder with PP (polypropylene matrix). They observed that SR and KA were greatly impacted by the percentage of neem wood saw particles, traverse speed and water jet pressure. Predicting the optimal surface roughness after abrasive water jet machining of the NFRP material (fill: the sundi wood dust) was the topic of the work of Jagadish et al. (2019). Based on the results and the methods applied (fuzzy logic and regression analysis), optimum input parameters and the fundamental impact of AMGS (abrasive material grain size) and AMFR (abrasive mass flow rate) on the resulting surface were evaluated. It also mentioned a finding that a significant amount of

crack propagation, some voids on the machined surface, were found due to the presence of moisture content in the natural fillers. The authors Jagadish and Gupta also processed the results in the form of a book publication: Abrasive Water Jet Machining of Engineering Materials, extended to the machining of metallic materials, polymer composites and ceramic composites. Hutyrová et al. (2016) dealt with verification of suitability of water jet/abrasive water jet application to disintegration of turning NFRP samples (wood particles + HDPE matrix). Surface quality was investigated by optical profilometry and the AWJ machining technology was evaluated as a suitable alternative to machining, subject to setting optimal process parameters [24,28–30].

This paper dealt with the evaluation of the characteristics of technological inheritance—namely, profile parameters of surface roughness (*Rz*) in relation to the changing input parameters of the AWJ cutting process. Surface quality of machined surfaces was analysed using an optical profilometer MicroProf FRT, and the macrostructure of the created surface was visualized using the VHX-6000 digital microscope at 50× magnification (Keyence, Osaka, Japan).

2. Materials and Methods

Material: in the experiment, samples with dimensions of 40 × 60 × 1500 mm were used (technology of the profile production: process of extrusion). The NFRP composite material consisted of high-density PE matrix filled with wood reinforcement in a ratio of 25/75 vol.% (+additives). The mechanical properties of the examined samples are expressed in the graphical display—Figure 2.

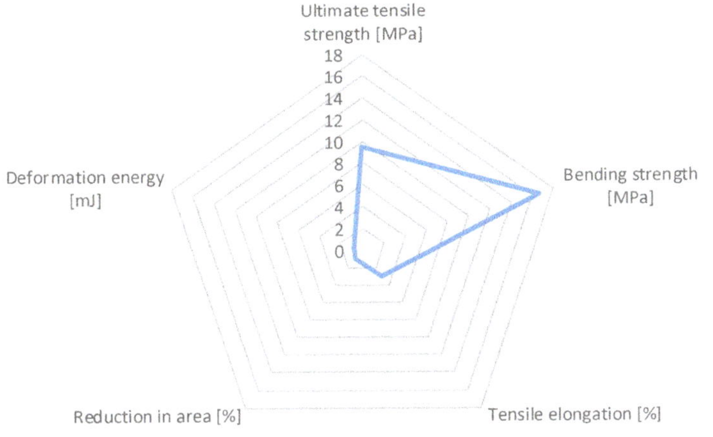

Figure 2. Mechanical properties of the examined wood plastic composite material (values are statistically verified).

Water Jet 3015 RT-3D was used for the AWJ cutting of the samples investigated. Required water pressure was supplied by the PTV JETS—3.8/60 Classic pump. Abrasive particles (Australian garnet, 80 MESH, were fed from a hopper through an abrasive management system by a tube with the inner diameter of 6.4 mm with the calibrated dosing accuracy of ±2.0 g. The two factors varied in course of the experiment (in relation to the controlled result parameter Q):

- Abrasive mass flow rate m_a [g·min^{-1}];
- Traverse speed of cutting head v_f [mm·min^{-1}].

The values of selected cutting process parameters (including sample designation) and machine settings are given in Tables 3 and 4.

Table 3. Design of sample testing and values of variable factors of the cutting process by AWJ.

Sample No.	Abrasive Mass Flow Rate m_a [g·min^{-1}]	Traverse Speed v_f [mm·min^{-1}]
S1	150	346
S2	200	387
S3	250	423
S4	300	455
↑	.	Q5
⋮	.	⋮
⋮	.	⋮
Q1	.	↓
S17	150	80
S18	200	90
S19	250	100
S20	300	105

The machined surfaces were scanned by the MicroProf FRT profilometer (working based on chromatic aberration), and the basic surface roughness characteristics were determined in accordance with ISO 21920-2, ISO 16610-21 (filtration of the selected parameters took place in accordance with the standard) [31,32]. The measurement was carried out in 20 lines along the sample, marked h1–h39, in direction perpendicular to direction of the water jet impact (Figure 3). Total length evaluated: 40 mm. Position of the last line < 2 mm from the lower edge—value determined by the SN 214001:2010 standard [33]. Figures 4 and 5 represents the lines which are considered as a transition between areas with different roughness.

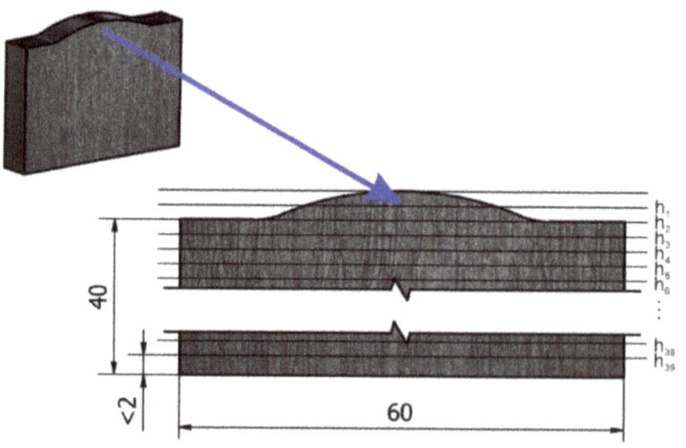

Figure 3. Positioning of lines along area under evaluation (last line least 2 mm from the bottom edge of the sample).

Table 4. Parameters of AWJ process cutting.

Parameters	Value	Unit
Water pressure p	400	MPa
Water orifice diameter d_0	0.3	mm
Focusing tube diameter d_f	0.9	mm
Standoff distance	4	mm
Abrasive mass flow rate m_a	150–300	g·min^{-1}
Traverse speed v_f	Variable [2]	mm·min^{-1}
Abrasive material	-	Australian Garnet
Abrasive particle	80	MESH

[2] Based on required cut quality Q1 to Q5 (set by the SN 214001:2010 standard [33]).

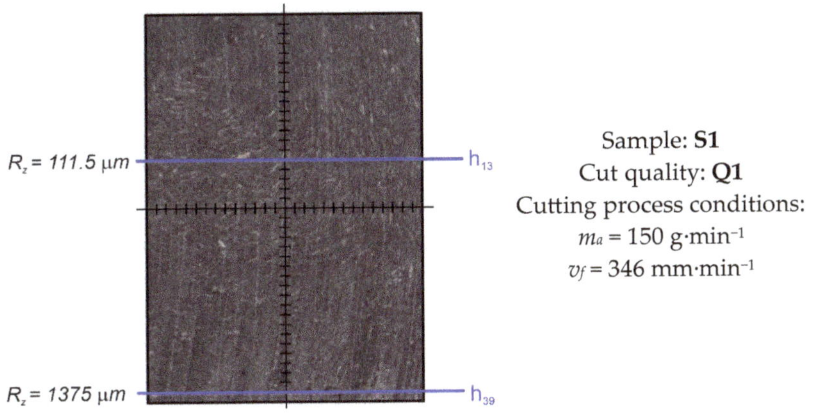

Figure 4. Selected lines h13/h39—measured of the *Rz* parameter in the lines defined (sample: S1, cut quality: Q1).

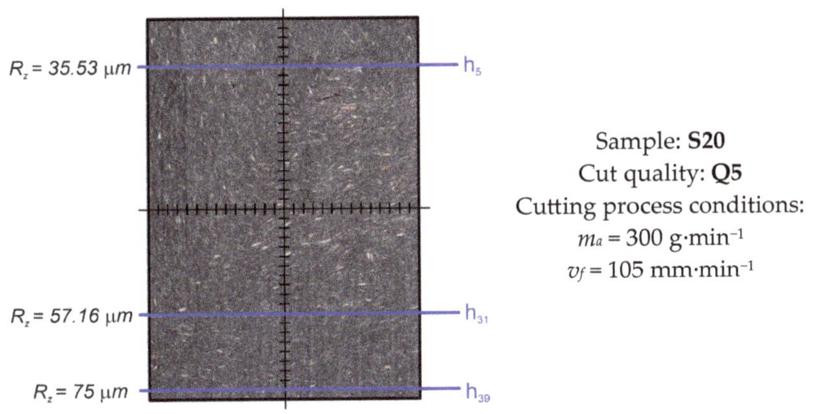

Figure 5. Selected lines h5/h31/h39—measured of the *Rz* parameter in the lines defined (sample: S20, cut quality: Q5).

Once the experimental measurements were performed and the measured data processed (by the Mountains SPIP 6.7.7 Academic software), information about the size of the surface topography unevenness dependent on the changing experimental technological parameters was obtained. In relation to the resulting Q parameter—it was necessary to evaluate individual sets separately. Based on SN 214001: 2010 (Contact-Free Cutting—Water Jet Cutting—Geometrical Product Specification and Quality from SAI Global,), the surfaces were categorized as Q1 to Q5, where the Q1 cut quality was defined as the lowest/Q5 cut quality as the highest [33].

3. Results and Discussion

In relation to the increasing trend of values, line charts were selected to display the Q1 and Q5 cut quality sets (making it possible to predict the *Rz* values between the lines recorded). Subsequently, the least squares method was used to construct the trend lines (regression equations—Table 5) of exponential dependencies (including the confidence factors) for the Q1 set. Based on the regression analysis of measured data, regression equations were created, which described dependence of maximal height of the profile (*Rz*) on the distance from the upper cut line. Since the trend of dependent variable was nonlinear, exponential equations were created after analysis. Exponential equation in basic format is represented by prescription:

$$Y = \beta_0 \cdot X^{\beta_1} + \varepsilon$$

where β_0 and β_1 are the model parameters and ε is the random observation error.

Table 5. Equations of exponential dependencies of the SET Q1.

Sample No.	Equation of the *Rz* Dependence in Relation to the Depth Line
S1	$y = 28.304e^{0.1914x} / R^2 = 0.9578$
S2	$y = 22.113e^{0.1931x} / R^2 = 0.9151$
S3	$y = 36.013e^{0.1748x} / R^2 = 0.9838$
S4	$y = 39.418e^{0.1529x} / R^2 = 0.9756$

Functional dependencies are considered statistically significant where $R^2 > 0.9$ (if the $\rightarrow$ exponential equation is $R^2 = 0$ not suitable for assuming the behaviour of values/$R^2 = 1$ $\rightarrow$ there is a perfect correlation between the actual values and the values specified by the trendline). In the case of Q1 set, all dependencies described were statistically significant. Based on microscopic observation, the surfaces of the Q1 set samples can be divided into 2 different zones: a smooth zone and a coarse zone. Significant plastic deformation of the surface was recorded starting from the line h13. The highest value of the *Rz* parameter measured was recorded in the sample S3 in the h39 line, of 1536.0 μm. With the increasing cut depth, the *Rz* parameter values increased exponentially (see Figure 6). The exponential trend of observed parameters can be explained, since, in the process of abrasive water jet cutting, loss of the kinetic energy of the water jet and abrasive particles occurred via lowering the velocity in the direction from upper to lower cut line.

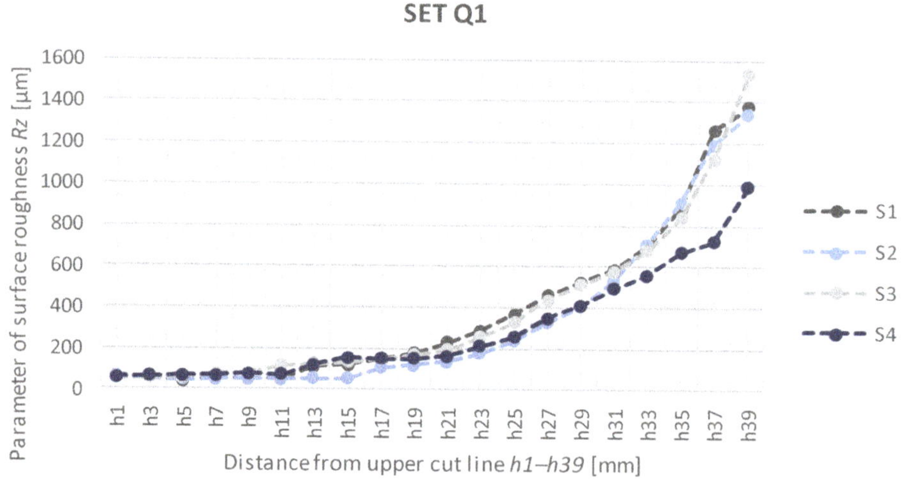

Figure 6. The course of the tallest height of *Rz* unevenness in lines h1–h39 (Q1 set samples).

In relation to the nature of the course of *Rz* parameters of the profile surface roughness, no regression dependencies for Q5 set were made (Figure 7). Samples of the Q5 set were divided into three zones: initiation, smooth (h5–h31) and coarse. The width of the initiation zone was defined by local minima in the h5 line (see Table 6). Within the defined smooth zone, the profile surface roughness parameters *Rz* ranged from 36.90 μm to 71.89 μm.

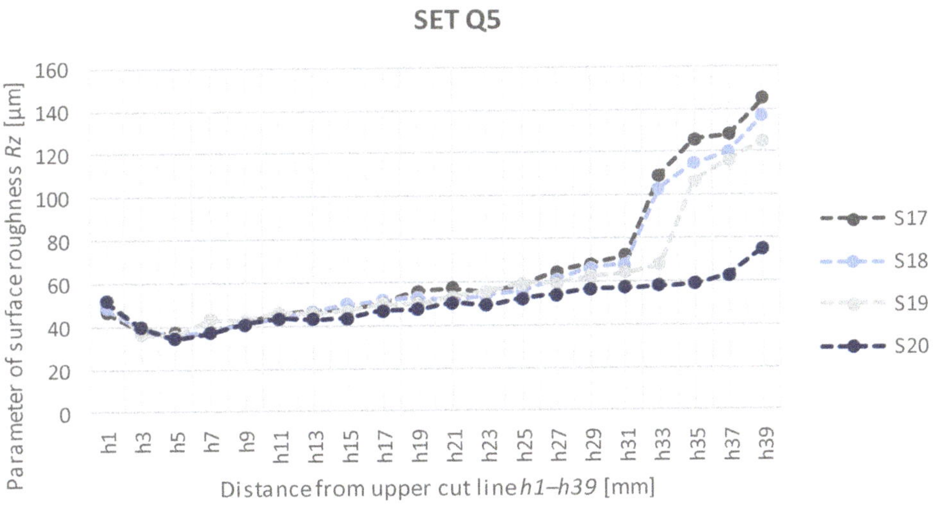

Figure 7. The course of the tallest height of Rz unevenness in lines h1–h39 (Q5 set samples).

Table 6. Rz parameter values in defined lines (defining transitions between individual zones).

Line Number *hx*	Sample S1 (Cut Quality: Q1)	Sample S2 (Cut Quality: Q1)	Sample S3 (Cut Quality: Q1)	Sample S4 (Cut Quality: Q1)
h13	111.5 µm	53.05 µm	123.50 µm	120.60 µm
h39	1375.0 µm	1341.0 µm	1536.0 µm	989.7 µm
Line Number *hx*	**Sample S17 (Cut Quality: Q5)**	**Sample S18 (Cut Quality: Q5)**	**Sample S19 (Cut Quality: Q5)**	**Sample S20 (Cut Quality: Q5)**
h5	37.00 µm	35.80 µm	35.67 µm	34.53 µm
h31	71.89 µm	68.14 µm	63.93 µm	57.16 µm
h39	144.70 µm	136.70 µm	125.00 µm	75.00 µm

The resulting surfaces of sets Q1 and Q5, and the associated unevenness, showed a specific morphology due to cutting under different technological parameters. Based on the shape on the lower cut line at the outlet of the water jet, it was possible to assess the "suitability" of the choice of technological parameters in relation to the physical and mechanical properties of the material [34].

The conditions proposed for set Q1 and set Q5 confirmed the previous claim. It was evident that with the decreasing traverse speed (and the increasing abrasive mass flow), the cut surfaces exhibited deformations that were less pronounced (Table 7, Figure 8). In Table 7, surfaces created by a defined combination of factors, abrasive mass flow rate m_a/traverse speed v_f, are visualized.

Figure 8. Cut at water jet outlet (left: Q1 set samples/right: Q5 set samples).

Table 7. Images of sample surface structure taken with the VHX-6000 digital microscope. Samples: S1–S4 (Q1), S17–S20 (Q5).

S1 (Cut Quality: Q1)	S2 (Cut Quality: Q1)	S3 (Cut Quality: Q1)	S4 (Cut Quality: Q1)

S17 (Cut Quality:Q5)	**S18 (Cut Quality: Q5)**	**S19 (Cut Quality: Q5)**	**S20 (Cut Quality: Q5)**

The macrostructure of the surface was visualized using the VHX-6000 digital microscope (Keyence) at 50× magnification. The surface area of 2.6 × 3.6 mm was scanned by the stapling method. In relation to the technology of profile production—by extrusion—the orientation of wood fibres and particles followed the flow of polymer. The photographed area (detail Figure 9) showed wood particles with different colours and fractions. At the interface of the components: wood versus plastic, cracks were visible that were created by imperfect encapsulation of wood particles (the cause of imperfect encapsulation is the low adhesion of wood particles to resins/matrices, the complex shape of the surface). Based on the analysis of samples from 1 to 20, their size ranged from a few micrometres to about 4 mm. The most pronounced cracks were on the samples designated 18 and 20. The mechanism of tracks (grooves/furrows) formation was given by the trajectory of movement and the effect of abrasive particles on the surface of the cut (Figure 10).

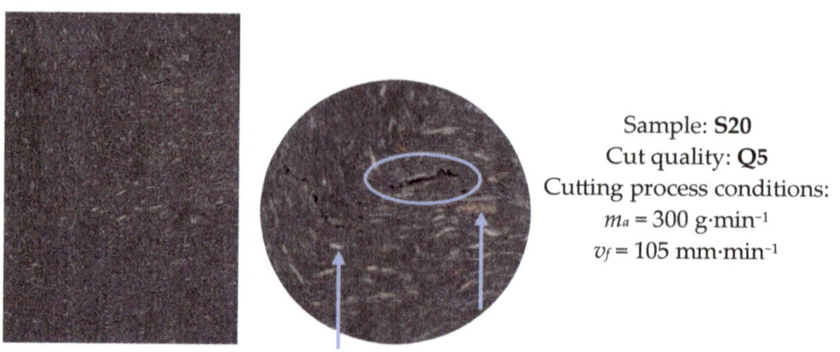

Figure 9. Significant cracks at the top of the profile cut by AWJ (arrows are pointing on natural reinforcement with different fractions and colours).

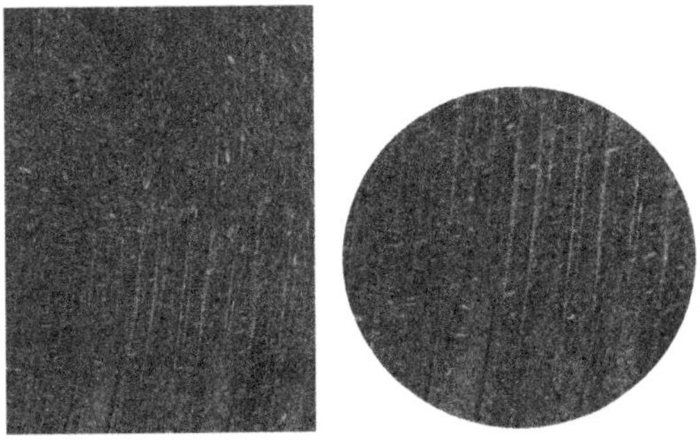

Figure 10. Pronounced furrows (tracks) left by AWJ in the final area of the cut.

4. Conclusions

Conventional machining of the composite materials with natural fibres is a long-lasting problem (compared to conventional metal working). The application of the AWJ technology eliminates the problem of tool wear, in the sense that the plastic matrix partially melts and sticks to the functional surfaces of the tool used. Surface topography was investigated by optical profilometry and macroscopic evaluation. Sets of samples with defined cut quality were analysed: Q1 (lowest cut quality) and Q5 (highest cut quality). The cutting parameters were selected with respect to the required quality Q (values of the ma parameter: selected/values of the v_f parameter: calculated by the machine control system with respect to the expected Q, based on the material library). It was evident that with the decreasing traverse speed of the cutting head (and increasing mass flow of the abrasive), the cut surfaces exhibited less pronounced deformations. In addition to different technological parameters of the sets, the resulting surface quality also depended on the physical properties of the material's NFRP components. Subject to optimal cutting process parameters, the AWJ technology can be applied as a suitable alternative for cutting the NFRP materials (considering the mechanical properties of the composite, the type and orientation of the fibres/particles).

Author Contributions: Conceptualization, Z.M.; methodology, D.K. and Z.M.; software, F.B. and R.V.; investigation, Z.M. and D.M.; resources, F.B.; data curation, D.K. and Z.M.; writing—original draft preparation, F.B. and Z.M.; writing—review and editing, F.B.; visualization, J.L.; supervision, Z.M. and F.B. All authors have read and agreed to the published version of the manuscript.

Funding: This work was supported by the Slovak Research and Development Agency under the contract No. APVV-20-0514 (Research of the influence of technological parameters of abrasive water jet machining on the surface integrity of tool steels). Research was supported by project KEGA no. 032TUKE4/2021 Transfer and implementation of knowledge in the field of energy beam technologies into study programs of technical secondary schools supporting dual education and project VEGA 1/0080/20 Research into the effect of high speed and high feed machining technologies on the surface integrity of hard-to-machine materials.

Data Availability Statement: Not applicable.

Conflicts of Interest: The authors declare no conflict of interest.

References

1. Hodzic, A.; Shanks, R. *Natural Fibre Composites: Materials, Processes and Properties*, 1st ed.; Woodhead Publishing: Sawston, UK, 2014; pp. 4–27.
2. Taiwo, E.M.; Yahya, K.; Haron, Z. Potential of using natural fiber for building acoustic absorber: A review. *J. Phys. Conf. Ser.* **2019**, *1262*, 012017. [CrossRef]
3. Hernandes Diat, D.; Villar-Ribera, R.; Haron, Z.; Julián, F.; Hernández-Abad, V.; Delgado-Aguilar, M. Impact Properties and Water Uptake Behavior of Old Newspaper Recycled Fibers-Reinforced Polypropylene Composites. *Materials* **2020**, *13*, 1079. [CrossRef] [PubMed]

4. Campilho, R.D.S.G. *Natural Fiber Composites*; Taylor & Francis Group: Boca Raton, FL, USA, 2016; p. 117.
5. Available online: https://www.baxcompany.com (accessed on 15 April 2023).
6. Kumar, S.P.; Sharma, S. Effect of Alkali Treatment on Jute Fibre Composites. Ph.D. Thesis, National Institute of Technology, Rourkela, India, 2007.
7. Lu, N.; Swan, J.R.; Ferguson, I. Composition, structure, and mechanical properties of hemp fiber reinforced composite with recycled high-density polyethylene matrix. *J. Compos. Mater.* **2012**, *46*, 1915–1924. [CrossRef]
8. Shahria, S. Fabrication and property evaluation of hemp-flax fiber reinforced hybrid composite. *Cellulose* **2019**, *7*, 17–23. [CrossRef]
9. Venkateshwaran, N.; Elayapemural, A. Mechanial and water absorption properties of woven jute/banana hybrid composites. *Fibers Polym.* **2012**, *13*, 907–914. [CrossRef]
10. Klyosov, A. Foreword-Overview: Wood-Plastic Composites. In *Wood-Plastic Composites*, 1st ed.; John Wiley and Sons: Hoboken, NJ, USA, 2007; pp. 42–45.
11. Chow, P.; Nakayama, F.S.; Youngquist, J.A.; Muehl, J.H.; Krzysik, A.M. Durability of wood/plastic composites made from Parthenium species. In *Thirty-Third Annual Meeting of the International Research Group on Wood Preservation*; IRG Secretariat: Stockholm, Sweden, 2002; pp. 2–11.
12. Stark, N.M.; Rowlands, R.E. Effect of wood fiber characteristics on mechanical properties of wood/polypropylene composites. *Wood Fiber Sci.* **2003**, *54*, 167–174.
13. Murayama, K.; Ueno, T.; Kobori, H.; Kojima, Y.; Suzuki, S.; Aoki, K.; Ito, H.; Ogoe, S.; Okamoto, M. Mechanical properties of wood/plastic composites formed using wood flour produced by wet ball-milling under various milling times and drying methods. *J. Wood Sci.* **2019**, *65*, 1–10. [CrossRef]
14. Schwarzkopf, M.J.; Burnard, N.D. Wood-plastic composites-Performance and environmental impacts. *Eviron. Impacts Tradit. Innov. For.-Based Bioprod.* **2016**, 19–43.
15. Mordor Intelligence. Available online: https://www.mordorintelligence.com/industry-reports/europe-wood-plastic-composite-market (accessed on 5 May 2023).
16. ENKEV. Available online: https://enkev.com/natural-products/cocolok (accessed on 6 May 2023).
17. Niu, S.; Chang, Q.; He, W.; Zhao, D.; Xie, Y.; Deng, X. Mechanically Strong, Hydrostable, and Biodegradable Strach-Cellulose Composite Materials for Tableware. *Starke* **2022**, *74*, 2200019. [CrossRef]
18. Chegdani, F.; El Mansori, M. New Multiscale Approach for Machining Analysis of Natural Fiber Reinforced Bio-Composites. *J. Manuf. Sci. Eng.* **2019**, *141*, 011004. [CrossRef]
19. Zajac, J.; Hutyrova, Z.; Orlovsky, I. Investigation of surface roughness after turning of one kind of the bio-materials with thermoplastic matrix and natural fibers. *Adv. Mater. Res.* **2014**, *941*, 275–279. [CrossRef]
20. Caggiano, A. Machining of fibre reinforced plastic composite materials. *Materials* **2018**, *11*, 442. [CrossRef] [PubMed]
21. Babu, D.; Babu, K.S.; Gowd, B. Drilling uni-directional fiber-reinforced plastics manufactured by hand lay-up influence of fibers. *JMST* **2012**, *1*, 1–12. [CrossRef]
22. Babu, D.; Babu, K.S.; Gowd, B. Determination of delamination and tensile strength of drilled natural fiber reinforced composites. *Appl. Mech. Mater.* **2014**, *592*, 134. [CrossRef]
23. Abilash, N.; Sivapragash, M. Optimizing the delamination failure in bamboo fiber reinforced polyester composite. *Jksues* **2016**, *28*, 92–102. [CrossRef]
24. Gupta, J. Abrasive Water Jet Machining of Polymer Composites. In *Abrasive Water Jet Machining of Engineering Materials*, 1st ed.; Springer: Cham, Switzerland, 2020; pp. 33–49.
25. Vigneshwaran, S.; Uthayakumar, M.; Arumugaprabu, V. Abrasive water jet machining of fiber-reinforced composite materials. *J. Reif. Plast Compos.* **2018**, *37*, 230–237. [CrossRef]
26. Masoud, F.; Sapuan, S.M.; Mohd Ariffin, M.K.A.; Nukman, Y.; Bayraktar, E. Cutting process of natural fiber-reinforced polymer composites. *Polymers* **2020**, *12*, 1332. [CrossRef]
27. Cuha, D.; Hatala, M. Effect of a modified impact angle of an ultrasonically generated pulsating water jet on aluminium alloy erosion using upward and downward stairs trajectory. *Wear* **2022**, *500*, 204369. [CrossRef]
28. Boopathi, S.; Thillaivanan, A.; Azeem, M.A.; Shanmugam, P.; Pramod, V.R. Experimental investigation on abrasive water jet machining of neem wood plastic composite. *Funct. Compos. Struct.* **2022**, *4*, 025001. [CrossRef]
29. Jagadish, S.B.; Amitava, R. Prediction of surface roughness quality of green abrasive water jet machining: A soft computing approach. *J. Intell. Manuf.* **2015**, *30*, 2965–2979. [CrossRef]
30. Hutyrova, Z. Turning of wood plastic composite by water jet and abrasive water jet. *Int. J. Adv. Manuf. Technol.* **2015**, *84*, 1615–1623. [CrossRef]
31. *ISO 21920-2:2021*; Geometrical Product Specifications (GPS)—Surface Texture: Profile—Part 2: Terms, Definitions and Surface Texture Parameters. ISO: Geneva, Switzerland, 2021.
32. *ISO 16610-21:2011*; Geometrical Product Specifications (GPS)—Filtration-Part 21: Linear Profile Filters: Gaussian Filters. ISO: Genava, Switzerland, 2011.

33. *SN 214001:2010*; Contact-Free Cutting—Water Jet Cutting—Geometrical Specification and Quality. Swiss Standards-Vereiningung: Zurich, Switzerland, 2013.
34. Hloch, S.; Valicek, H. *Vplv Faktorov na Topografiu Povrchov Vytvorených Hydroabrazívnym Delením*, 1st ed.; SH: Presov, Slovakia, 2008; pp. 66–99.

machines

Article

Artificial Neural Network-Based Predictive Model for Finite Element Analysis of Additive-Manufactured Components

Sorin D. Grozav [1], Alexandru D. Sterca [1,*], Marek Kočiško [2], Martin Pollák [2] and Vasile Ceclan [1]

1 Faculty of Industrial Engineering, Robotics and Production Management, Technical University of Cluj-Napoca, Bulevardul Muncii nr. 103-105, 400641 Cluj-Napoca, Romania; sorin.grozav@tcm.utcluj.ro (S.D.G.); vasile.ceclan@tcm.utcluj.ro (V.C.)
2 Faculty of Manufacturing Technologies with a Seat in Presov, Technical University of Kosice, Bayerova 1, 080 01 Presov, Slovakia; marek.kocisko@tuke.sk (M.K.); martin.pollak@tuke.sk (M.P.)
* Correspondence: alexandru.sterca@iirmp.utcluj.ro; Tel.: +40-752141875

Abstract: Additive manufacturing is becoming one of the most utilized tools in an increasing number of fields from Industry 4.0 concepts, engineering, and manufacturing to aerospace and medical applications. One important issue with additive-manufactured components is their orthotropic behaviour where mechanical properties are concerned. This behaviour is due to the layer-by-layer manufacturing process and is particularly hard to predict since it depends on a number of factors, including the manufacturing parameters used during the manufacturing process (speed, temperature, etc.). This study aimed to create and train an artificial neural network-based predictive model using empirical tensile strength data obtained from additive manufactured test parts using the FDM method and PLA material. The predictive model was designed to predict mechanical characteristics for different orientation axis, which were used to set the material properties for finite element analysis. Results indicate a strong correlation between predicted finite element analysis behaviour and real-world tests on additive-manufactured components. The neural network model was trained to an accuracy of ~93% for predicting the mechanical characteristics of 3D-printed PLA material. Using the predicted mechanical characteristics for defining a custom orthotropic material profile in finite element analysis, the simulated failure mode and the behaviour of a complex geometry component agreed with the real-world test.

Keywords: machine learning; finite element analysis; artificial neural network; additive manufacturing; high-precision metrology; CAD; predictive model

1. Introduction

The advantages of additive manufacturing technologies make them an attractive option in a variety of fields. Direct CAD to finished product manufacturing, the ability to fabricate components of high geometric complexity, and the overall flexibility and versatility of additive manufacturing techniques have contributed to an increase in the exploitation of these technologies. For all the advantages, additive manufacturing also suffers from a series of disadvantages, for example, low dimensional accuracy and surface quality. One of the most limiting factors of additive manufacturing is the orthotropic mechanical behaviour of the components fabricated using these technologies. Due to the layer-by-layer fabrication process intrinsic to additive manufacturing, the parts obtained present different mechanical properties for different orientation planes. As such, any given additive manufactured component tends to have a lower tensile strength in the direction perpendicular to the planes of the deposited layers. The tensile strength of additive manufactured components in the direction parallel to the deposited layers can also differ drastically from the tensile strength of the bulk material, mostly because the walls and, thus, the contours of the fabricated parts are generated by lines which are fused by sintering (in the case of selective laser sintering), melting (for fused deposition modelling and selective laser melting), or polymerisation (in the case of SLA).

Due to the nature of additive manufacturing technologies and the strategy of fusing lines to form walls and layers to build up the parts, the mechanical behaviour of additive-manufactured components is hard to model since it depends on a large number of factors, for example, the different process parameters (manufacturing speed, temperature, part orientation, etc.) and interactions between factors.

Previous studies by A.D. Sterca et al. [1] determined by statistical methods that there is a correlation between FDM process parameters and the dimensional accuracy of the printed parts. Further studies by S.D. Grozav et al. [2] tested, with good results, the feasibility of employing neural-network-based models for predicting the dimensional accuracy as well as the mechanical properties of components obtained by FDM additive manufacturing.

Work by Rodriguez et al. [3] used statistical methods with good results to determine the influence of parameters such as nozzle diameter, layer height, infill density, printing speed, layer orientation, and infill pattern on the mechanical characteristics of components fabricated using FDM from PLA material.

A study by Tura et al. [4] used predictive models to determine, to a good degree of accuracy, the effects of parameters such as raster angle, orientation angle, air gap, raster width, and layer height on the mechanical properties of FDM-manufactured parts using ABS material. Studies by Yadav D. et al. [5] were carried out on improving the tensile strength of additive manufactured parts by employing neural networks trained on parameters such as infill density, extrusion temperature, and material density, showing a 4.54% improvement in subsequent samples. Saleh et al. [6] performed a study using mathematical and neural-network-based predictive models to determine the mechanical characteristics of lattice structures fabricated using carbon fibre/PLA material, reaching a prediction accuracy of ~93%. Jatti et al. [7] performed research on optimising the tensile strength of FDM-manufactured components by using machine learning algorithms and the desirability approach with great results, showing a good correlation between predictions and real-world data. Ding et al. [8] conducted studies on special plastics for engineering, e.g., PEI and PEEK, showing a strong correlation between process parameters and mechanical behaviour, allowing them to optimize the mechanical characteristics by adjusting the process parameters. Studies [9] performed on components printed using different orientation angles were also performed by Khosravani et al., evidentiating the orthotropic behaviour of additive-manufactured components. Zhang et al. [10] performed a study on reducing dimensional error and warping of 3D-printed components by optimizing the process parameters using the Taguchi method and fuzzy comprehensive evaluation, showing that the most influential factor in the study is the extrusion speed of the material. Alhazmi et al. [11] performed a study on the influence of raster angle and infill density, showing different mechanical behaviour and failure modes for different combinations of parameters. Omiyale et al. [12] studied the effects of process parameters using the Taguchi method, predicting combinations of parameters that can be optimized for different mechanical characteristics of the 3D-printed components. Tran et al. [13] researched the possibility of improving 3D-printed components' quality and mechanical behaviour by varying the process parameters according to experimental data and results reported in the literature, showing that the components produced using the recommended parameters presented good dimensional and geometric characteristics as well as good mechanical behaviour. Peng et al. [14] applied genetic algorithms for the prediction of optimized process parameters for the manufacturing of 3D-printed components, showing an increase in part quality. Chacon [15] studied the effects of process parameters on the mechanical properties of 3D-printed structures, leading to a mathematical model for the selection of the optimal combination of parameters for a given geometry. Ayatollahi et al. [16] performed an in-depth analysis of the mechanical properties of 3D-printed parts using advanced techniques such as scanning electron microscopy, showing and explaining the different failure modes of the test samples.

As can be seen from these studies, the behaviour of 3D-printed components can be characterized using predictive models and statistical techniques with varying degrees of accuracy, encouraging further research in the field.

The study presented in this paper attempted to address the issue of modelling the mechanical behaviour of additive-manufactured components by employing an artificial neural network-based predictive model to determine the mechanical characteristics of 3D-printed parts which can be used to inform material profiles for use in finite element analysis, enabling the simulation of the behaviour of these components under load.

The study was performed on a set of 48 test samples fabricated using fused deposition modelling (FDM) with different parameters and combinations of parameters (nozzle temperature, printing speed, orientation). The material used for testing was PLA (polylactic acid), and all the test samples were fabricated from the same roll of material. The fabricated components were subjected to tensile strength measurements, providing empirical data on the mechanical characteristics of different process parameter combinations. PLA was chosen because in a previous study [2] it was shown to have the highest variability of mechanical properties for different process parameter combinations and because it is one of the most popular 3D-printing materials. Another reason for choosing PLA is the availability of more data from previous studies on components manufactured using this material. These data can be used to average out the noise and outliers in the measurements, increasing the statistical significance of the data and the accuracy of the predictive mathematical model.

In order to predict the mechanical characteristics of a given set of process parameters, a predictive model needs to be devised. Because mechanical characteristics of the 3D-printed components depend on the bonding strength between both the layers and the lines which form the wall, the predictive model needs to be able to approximate very complex functions; therefore, the model used in this study employed an artificial neural network which was trained on the empirical data obtained from tensile strength measurements of the test samples. The predicted mechanical characteristics, together with empirical data obtained from strain–stress measurements, were used to create an orthotropic material profile for finite element analysis, which can be used for modelling the behaviour of complex additive manufactured components.

2. Materials and Methods

2.1. Test Sample Preparation

The study was performed on 48 test samples which were manufactured according to the ISO527-2 standard [17] using the FDM additive manufacturing technique. The test sample used with relevant dimensions for the ISO527-1B tensile test sample can be seen in Figure 1.

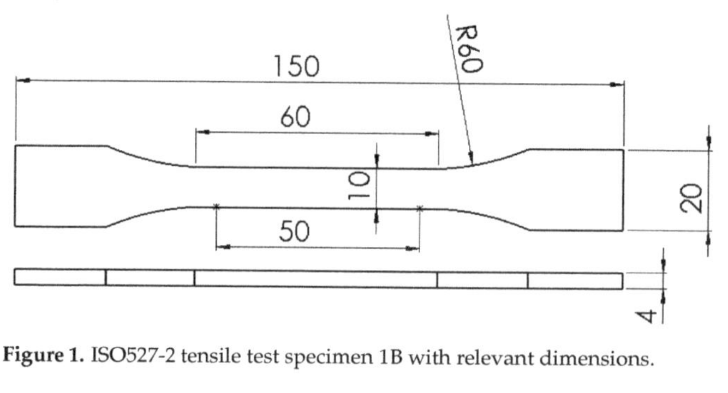

Figure 1. ISO527-2 tensile test specimen 1B with relevant dimensions.

Manufacturing of the test samples was performed in pairs of two, one horizontal and one vertical, positioned in the central area of the machine build platform, ensuring the same manufacturing conditions in both orientations. The process parameters (speed and nozzle temperature) were varied according to Table 1.

Table 1. FDM process parameter combinations used to manufacture the tensile test specimen pairs.

Speed (mm/s)	30	60	90	120	150	180
Nozzle temperature (°C)	190	190	190	190	190	190
	200	200	200	200	200	200
	210	210	210	210	210	210
	220	220	220	220	220	220

The fabrication of the test samples was performed using an Anycubic Mega X 3D printer [18], which is presented in Figure 2. The test samples were manufactured using a 0.4 mm diameter nozzle at a layer height of 0.2 mm. The infill was 100%; however, a strategy of printing the samples using only concentric walls instead of an infill pattern was chosen, to further evidentiate the orthotropic behaviour.

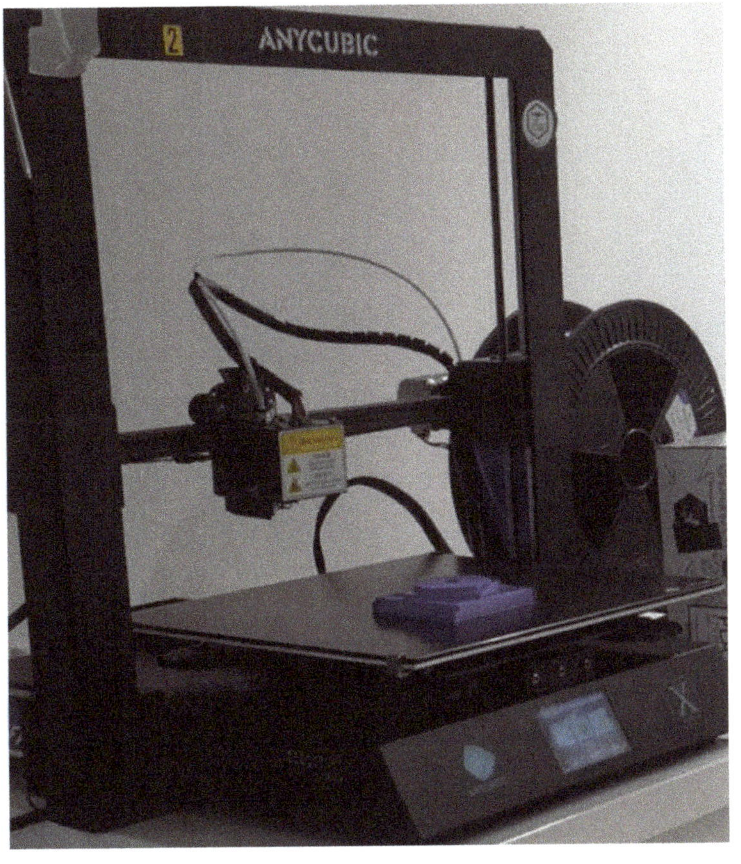

Figure 2. Anycubic Mega X 3D printer used for the manufacturing of the tensile test specimens.

An image of the manufactured tensile test specimens is presented in Figure 3a, while an image of the test specimens after tensile testing is presented in Figure 3b. The principal physical and mechanical characteristics of the PLA material in bulk are presented in Table 2.

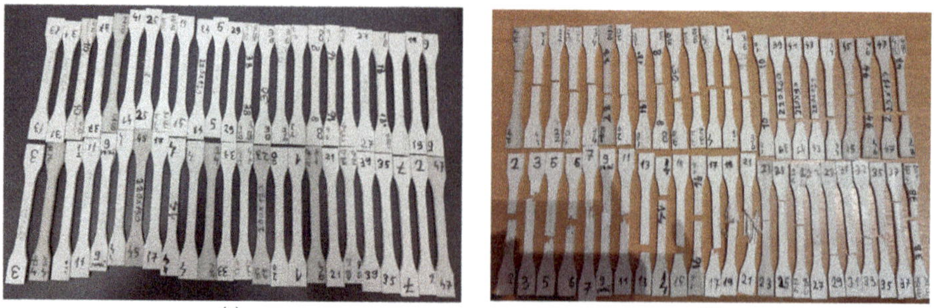

(**a**) (**b**)

Figure 3. Manufactured tensile test specimens (**a**); test specimens after tensile testing (**b**).

Table 2. Mechanical and physical characteristics of bulk PLA.

Property	Value
Heat Deflection Temperature (HDT)	126 °F (52 °C)
Density	1.24 g/cm^3
Tensile Strength	50 MPa
Flexural Strength	80 MPa
Impact Strength (Unnotched) IZOD (J/m)	96.1
Shrink Rate	0.37–0.41% (0.0037–0.0041 in/in)

2.2. Tensile Testing

Tensile testing of the samples was performed on an Instron 3366 machine [19] seen in Figure 4. The test is performed on a measuring length of 50 mm at the speed of 6.6 m/s and provides both tensile strength measurements and strain measurements, which are needed for training the predictive model and creating the material profile for finite element analysis.

Figure 4. Instron 3366 tensile testing machine.

The test parts after tensile testing are presented in Figure 3b, where the failure mode can be observed.

2.3. Neural Network Development and Training

The predictive model has at its core an artificial neural network which takes process parameters (speed, temperature, orientation) as inputs and empirical data in the form of tensile strength from tensile testing as outputs for the training process. The neural network architecture choice was informed by previous studies [2], where a number of neural network architectures were developed and tested for performance and compared to statistical techniques and models from other studies [1]. The presented neural network was tested against the best model from the study [2] as a reference point, and the parameters and architecture were adjusted to increase prediction accuracy.

The neural network was developed in the Python [20] programming language using the PyCharm community edition IDE [21]. In order to ensure cross-platform compatibility and to make use of efficient machine learning algorithms and functions, the Tensorflow [22] machine learning library was employed with the Keras [23] front-end.

Performance analysis and graphing for the neural network were performed using the Tensorboard [24] module from Tensorflow.

The architecture of the neural network employed in the study consisted of three fully connected hidden layers containing eight nodes each. For the inputs and outputs of the neural network, there were two possible configurations, one where the layer orientation parameter was used as a categorical input variable, and the output of the neural network consisted of one node for tensile strength values. The second configuration took as inputs the temperature, speed, and intrinsic tensile strength of the material, while the output consisted of two nodes, one for tensile strength in the horizontal orientation and the other in the vertical orientation, providing predictions for both orientations given a set of process parameters. Both architectures were tested; however, the first configuration was preferred since it allows training on uneven datasets where the number of data points for one orientation may differ from the other. A graphical representation of the neural network architecture for both configurations is presented in Figure 5.

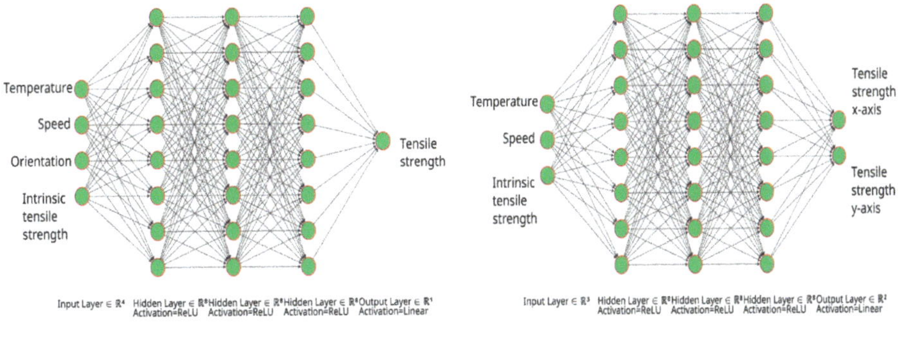

(**a**) (**b**)

Figure 5. Artificial neural network architecture: single-output configuration (**a**), and two-output configuration (**b**).

The inputs for the neural network consisted of the main process parameters, printing speed, nozzle temperature, and the intrinsic tensile strength of the bulk material. The intrinsic tensile strength of the bulk material is relevant only if the neural network is trained on data obtained from test samples fabricated from a variety of materials, which is the subject of a future study. The bulk material tensile strength can be obtained by subjecting a section of filament to tensile strength measurements. In the case of this study, all test samples were manufactured from the same roll of filament; therefore, the bulk material tensile strength was the same for all components, in which case, the neural network will ignore it or normalize it to 1 during training.

The outputs of the neural network use empirical tensile test data during training and provide predicted tensile strength values in MPa during inference.

The optimizer [25] used in the neural network was Adamax [26] and the loss function was MeanAbsolutePergentageError. The neural network weights and biases were initialized as zeroes. The activation function [27] for the hidden layer was ReLU (rectified linear unit), and for the output layer, a linear activation function was implemented. The activation function was chosen based on performance evaluation and the input/output data ranges. Performance evaluations of neural networks developed for this study and previous work [2] show better generalization and convergence as well as increased accuracy when using the ReLU activation function for hidden layers and a linear function for the output layers. This may be due to the range of the input and output variables not being constrained between a min and max value.

The tensile test data were split into 80% for training and 20% for validation. The percentages of data used for training and validation were chosen to optimize the training on a sparse dataset with a limited number of input and output values; reducing the percentage of available training data could lead to overfitting of the model during sparse-data scenarios. The validation data are never seen by the neural network during training, which ensures that the model can predict real-world values, giving an unbiased evaluation of the neural network performance. The neural network was trained until the mean absolute percentage error for the validation predictions reached the lowest point (highest accuracy) and the model with the best accuracy was saved.

2.4. Finite Element Analysis Using Predicted Mechanical Properties

The tensile strength predictions obtained from the neural network can be used together with empirical data from tensile testing to define a custom orthotropic material profile. By applying the custom material profile to the CAD model of a functional component, finite element analysis can be employed to determine the behaviour of the part during the design stage. Finite element analysis was performed using the simulation module available in the Solidworks software suite [28]. The material profile definition using values predicted by the artificial neural network is presented in Figure 6.

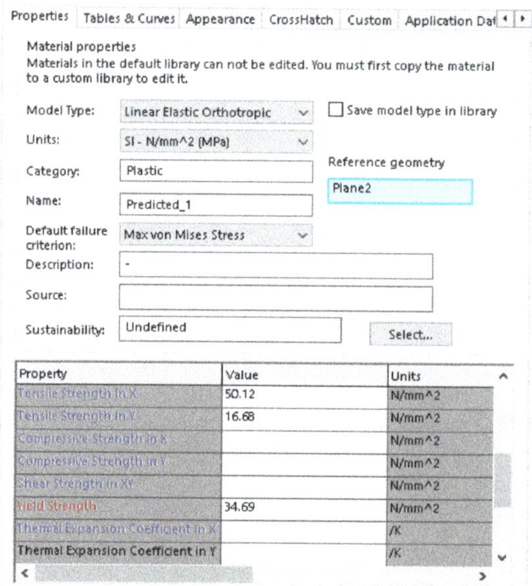

Figure 6. Orthotropic material definition in Solidworks using predicted mechanical characteristics.

2.5. Corroborating Predicted Results with Real-World Data

Finite element analysis was performed on the CAD model of a part which was fabricated using the FDM method and subjected to tensile testing. A blended-curvature type mesh was applied to the part with a maximum element size of 3 mm and minimum element size of 0.15 mm, containing 12,443 elements and 22,236 nodes. The external load was 1000 N applied on the internal surface of the 15 mm diameter holes, in opposite directions, simulating a tensile strength test. The average simulation time was ~30 s. The predicted behaviour and tensile strength were compared to the observed real-world test.

The CAD model of the part used for testing is shown in Figure 7. The geometry is the same as for the components used for previous studies, allowing the use of previous data to further evaluate the predictive model.

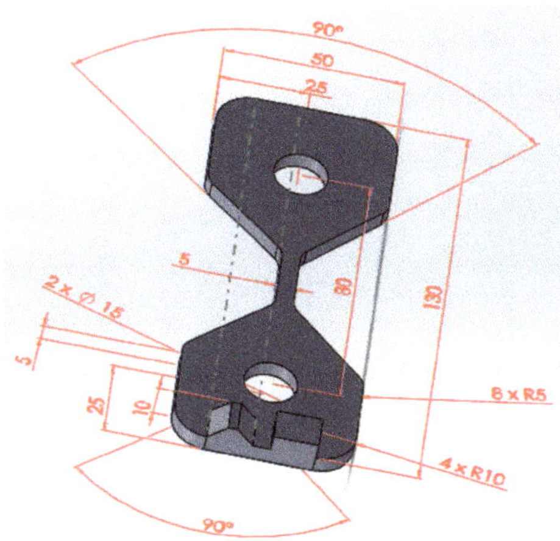

Figure 7. CAD model of the test part used for analysis.

Because the failure mode of 3D printing components is dependent on factors such as infill pattern and density, wall line count, and top/bottom thickness, the CAD model used for finite element analysis was designed to include these features, as can be observed in the section presented in Figure 8a.

(a) (b)

Figure 8. Section through the CAD model of the test part (**a**) and manufactured test part (**b**).

The material for the test part was chosen to have the same mechanical characteristics as the material for the test samples, with a tensile strength of ~60 Mpa (average over 10 measurements). The process parameters used for fabricating the test part were 195 °C nozzle temperature, 45 mm/s printing speed, and horizontal orientation. An image of the manufactured test part is presented in Figure 8b.

3. Results

3.1. Tensile Test Results

The tensile test results for the test samples are displayed in Figure 9. The graph indicates that the tensile strength is much lower in the direction perpendicular to the layer orientation, due to it being determined by the bonding strength between the layers, which is in turn affected by the temperature of the deposited material and the printing speed.

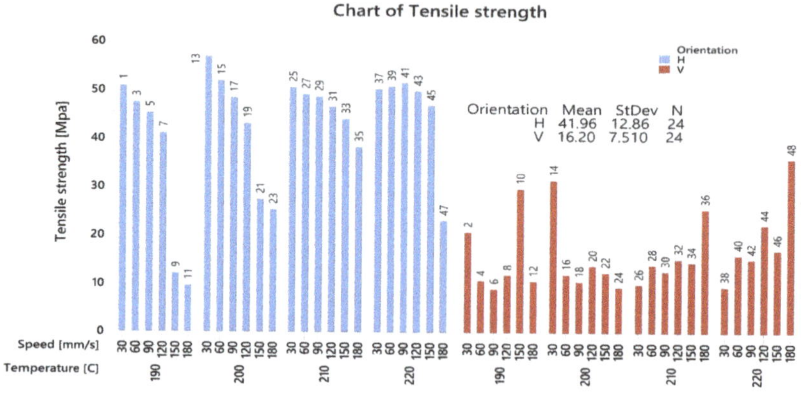

Figure 9. Results of tensile strength tests on specimens.

3.2. Neural Network Training and Prediction Results

The tensile test results were used as empirical data to train the neural network which will provide predictions for future parameter combinations. Figure 10 shows a performance plot for the trained neural network in the form of training data versus validation data prediction mean absolute percentage error. The results show a prediction accuracy of ~93%, indicating a strong correlation between process parameters and tensile strength and the ability of the neural network to approximate the complex function needed to predict tensile strength values for a given set of parameters.

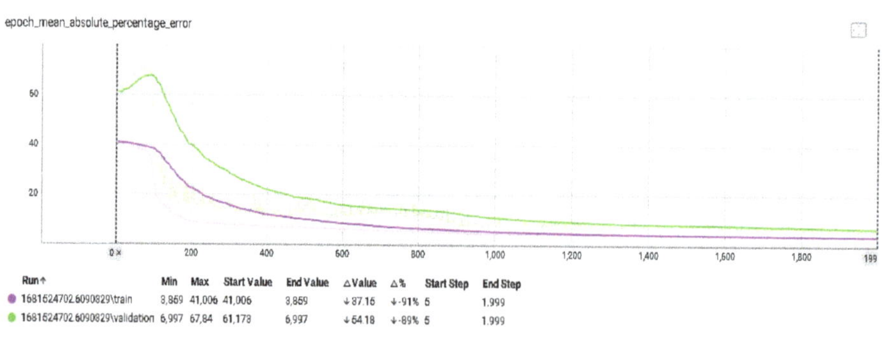

Figure 10. Performance analysis plot for the artificial neural network used to predict tensile strength values.

The predicted tensile strength value for the set of parameters used to manufacture the test part (temperature 195 °C, speed 45 mm/s horizontal orientation) was 50.12 MPa.

3.3. Finite Element Analysis Results

The predicted values obtained from the neural network were used to create a custom material with orthotropic properties for finite element analysis. A test part was manufactured with the same process parameters used for making the prediction and was subjected to tensile testing. The results of the finite element analysis are shown in Figure 11.

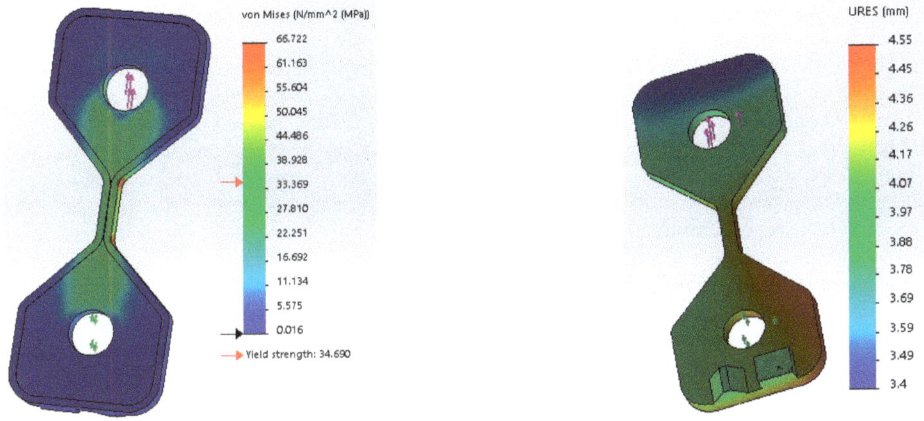

(a) (b)

Figure 11. Finite element analysis results for orthotropic material with predicted mechanical characteristics. Tensile stress results (**a**); displacement results (**b**).

For comparison, the results of the tensile test for the manufactured test part are shown in Figure 12, where (a) is an image of the part after breaking and (b) is the stress–strain plot generated while the part was under test.

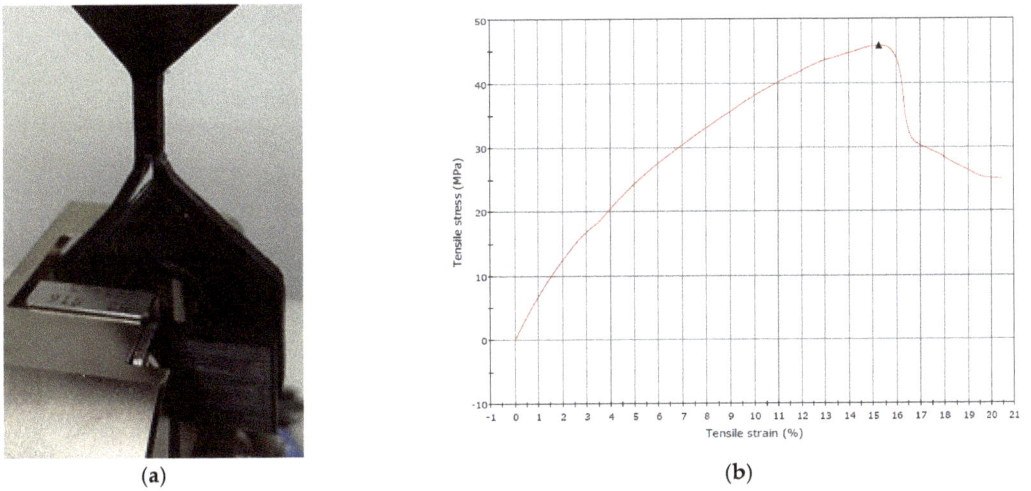

(a) (b)

Figure 12. (**a**) Test part failure mode under tensile testing. (**b**) Stress–strain plot generated during tensile testing of the part, indicating maximum tensile stress (upward triangle).

The test part failure mode shows that the contour walls broke away from the top and bottom layers, indicating low bonding strength between them and a weak point for the mechanical behaviour of the part. The same areas were identified by finite element analysis as weak points with a high probability of failure under load. The results indicate that the model can predict the behaviour of FDM-manufactured components for both tensile strength and failure mode.

4. Discussion

The findings agree with results from other studies in the literature, with studies performed by Khosravani et al. [9], Alhazmi et al. [11], and others showing the same strong relationship between part build orientation and mechanical characteristics. Studies by Ding et al. also showed a strong correlation between orientation and mechanical characteristics, and at the same time, an influence of nozzle temperature was observed. Many of the studies in the literature used statistical methods for characterizing the influence of parameters and predicting mechanical behaviour; however, the results of this study and research performed by Yadev et al. [5] show that artificial neural networks, due to their ability to approximate complex functions, can provide higher accuracy predictions.

The results show that the simulated and actual behaviours are similar to within an acceptable margin of error for the test part and the process parameters analysed. This indicates that artificial neural network-based predictive models can be used to augment finite element analysis by predicting the mechanical characteristics of materials in the context of FDM manufacturing.

A degree of error in the predicted and simulated data is present and to be expected; however, the accuracy of the predictions and the results are sufficient to prove the concept. The source of the errors is the limited number of data points available and the noise which is to be expected in a limited dataset. Another source of errors can be factors that influence the mechanical behaviour of FDM-fabricated parts but were not included as parameters for training the neural network. The inclusion of these factors (layer cooling, ambient temperature, print bed temperature, etc.) may be the subject of a future study.

This study builds on and validates the results of previous studies carried out in the field, with each study adding to the number of measurement data points leading to an increase in the precision and complexity of predictive models, encouraging future research in the field.

Future research is proposed to include more materials and process parameter combinations as well as a more extensive dataset which helps in reducing the number of outliers in the analysed data by averaging out the noise and increasing the precision and the versatility of predictive models. The results are promising enough to warrant future research on implementing neural-network-based models for predicting the mechanical behaviour of materials in the context of other manufacturing technologies such as SLS, SLM, and SLA.

5. Conclusions

This study shows that using artificial neural networks to predict orthotropic, FDM, 3D-printed material mechanical characteristics can aid finite element analysis in simulating the mechanical behaviour of 3D-printed components.

This kind of predictive model is useful for reducing the costs of additive manufacturing by predicting the mechanical behaviour of 3D-printed components during the design phase and informing decisions on either redesigning the part or choosing different process parameters.

Future research on more extensive datasets can increase the accuracy, complexity, and versatility of the predictions.

Author Contributions: Conceptualization, S.D.G., A.D.S. and V.C.; methodology, S.D.G., A.D.S. and V.C.; software, A.D.S.; validation, S.D.G. and V.C.; formal analysis, S.D.G., A.D.S. and V.C.; investigation, S.D.G., A.D.S. and V.C.; resources, S.D.G. and V.C.; data curation, A.D.S.; writing—original draft preparation, A.D.S.; writing—review and editing, M.K. and M.P.; visualization, A.D.S.; supervision,

S.D.G.; project administration, V.C.; funding acquisition, S.D.G. All authors have read and agreed to the published version of the manuscript.

Funding: This research was funded by the Technical University Of Cluj-Napoca, and the APC was funded by the Technical University Of Cluj-Napoca.

Acknowledgments: This paper has been elaborated in the framework of the project VEGA 1/0121/23.

Conflicts of Interest: The authors declare no conflict of interest.

References

1. Sterca, A.D.; Calin, R.A.; Cristian, L.; Walcher, E.M.; Bodur, O.; Ceclan, V.; Grozav, S.D.; Durakbasa, N.M. Evaluation of Fused Deposition Modeling Process Parameters Influence on 3D Printed Components by High Precision Metrology. In *Digitizing Production Systems;* Lecture Notes in Mechanical Engineering; Durakbasa, N.M., Gençyılmaz, M.G., Eds.; Springer: Cham, Switzerland, 2022.
2. Grozav, S.D.; Sterca, A.D.; Kočiško, M.; Pollák, M.; Ceclan, V. Feasibility of Predictive Models for the Quality of Additive Manufactured Components Based on Artificial Neural Networks. *Machines* **2022**, *10*, 128. [CrossRef]
3. Travieso-Rodriguez, J.A.; Jerez-Mesa, R.; Llumà, J.; Traver-Ramos, O.; Gomez-Gras, G.; Roa Rovira, J.J. Mechanical Properties of 3D-Printing Polylactic Acid Parts subjected to Bending Stress and Fatigue Testing. *Materials* **2019**, *12*, 3859. [CrossRef] [PubMed]
4. Tura, A.D.; Lemu, H.G.; Mamo, H.B. Experimental Investigation and Prediction of Mechanical Properties in a Fused Deposition Modeling Process. *Crystals* **2022**, *12*, 844. [CrossRef]
5. Yadav, D.; Chhabra, D.; Garg, R.K.; Ahlawat, A.; Phogat, A. Optimization of FDM 3D printing process parameters for multi-material using artificial neural network. *Mater. Today Proc.* **2020**, *21*, 1583–1591. [CrossRef]
6. Saleh, M.; Anwar, S.; Al-Ahmari, A.M.; AlFaify, A.Y. Prediction of Mechanical Properties for Carbon fiber/PLA Composite Lattice Structures Using Mathematical and ANFIS Models. *Polymers* **2023**, *15*, 1720. [CrossRef] [PubMed]
7. Jatti, V.S.; Sapre, M.S.; Jatti, A.V.; Khedkar, N.K.; Jatti, V.S. Mechanical Properties of 3D-Printed Components Using Fused Deposition Modeling: Optimization Using the Desirability Approach and Machine Learning Regressor. *Appl. Syst. Innov.* **2022**, *5*, 112. [CrossRef]
8. Ding, S.; Zou, B.; Wang, P.; Ding, H. Effects of nozzle temperature and building orientation on mechanical properties and microstructure of PEEK and PEI printed by 3D-FDM. *Polym. Test.* **2019**, *78*, 105948. [CrossRef]
9. Khosravani, M.R.; Reinicke, T. Effects of raster layup and printing speed on strength of 3D-printed structural components. In Proceedings of the 1st European-Structural-Integrity-Society (ESIS) Virtual European Conference on Fracture (ECF), Virtual, 29 June–1 July 2020; pp. 720–725.
10. Zhang, J.W.; Peng, A.H. Process-parameter optimization for fused deposition modeling based on Taguchi method. *Adv. Mater. Res.* **2012**, *538*, 444–447. [CrossRef]
11. Alhazmi, M.W.; Backar, A.; Backar, A.H. Influence of infill density and orientation on the mechanical response of PLA+ specimens produced using FDM 3D printing. *Int. J. Adv. Sci. Technol.* **2020**, *29*, 3362–3371.
12. Omiyale, P.K.; Omiyale, B.O. Mechanical behaviour of polylactic acid parts fabricated via material extrusion process: A taguchi-grey relational analysis approach. *Int. J. Eng. Res. Afr.* **2020**, *46*, 32–44.
13. Tran, N.H.; Nguyen, V.N.; Ngo, A.V.; Nguyen, V.C. Study on the effect of fused deposition modeling (FDM) process parameters on the printed part quality. *Int. J. Eng. Res. Appl.* **2017**, *7*, 71–77.
14. Peng, A.; Xiao, X.; Yue, R. Process parameter optimization for fused deposition modeling using response surface methodology combined with fuzzy inference system. *Int. J. Adv. Manuf. Technol.* **2014**, *73*, 87–100. [CrossRef]
15. Chacon, J.M.; Caminero, M.A.; Garcia-Plaza, E.; Nunez, P.J. Additive manufacturing of PLA structures using fused deposition modelling: Effect of process parameters on mechanical properties and their optimal selection. *Mater. Des.* **2017**, *124*, 143–157. [CrossRef]
16. Ayatollahi, M.R.; Nabavi-Kivi, A.; Bahrami, B.; Yahya, M.Y.; Khosravani, M.R. The influence of in-plane raster angle on tensile and fracture strengths of 3D-printed PLA specimens. *Eng. Fract. Mech.* **2020**, *237*, 106905. [CrossRef]
17. *ISO527-2*; Plastics-Determination of Tensile Properties-Part 2: Test Conditions for Moulding and Extrusion Plastics. International Organization for Standardization: Geneva, Switzerland, 2012.
18. Anycubic Mega, X. Anycubic, China. Available online: https://de.anycubic.com/products/mega-x (accessed on 15 April 2023).
19. Instron 3360 Series Mechanical Testing Systems. Available online: https://www.instron.com/~/media/literature-library/products/2011/06/3300-series-table-model.pdf (accessed on 27 December 2021).
20. Python Programming Language. Available online: https://www.python.org/ (accessed on 27 December 2021).
21. PyCharm Integrated Development Environment. Available online: https://www.jetbrains.com/pycharm/ (accessed on 27 December 2021).
22. TensorFlow Machine Learning Platform. Available online: https://www.tensorflow.org/ (accessed on 27 December 2021).
23. Keras Frontend for TensorFlow. Available online: https://keras.io/ (accessed on 27 December 2021).
24. TensorBoard Visualisation Kit for TensorFlow. Available online: https://www.tensorflow.org/tensorboard (accessed on 27 December 2021).

25. An Overview of Gradient Descent Optimization Algorithms. Available online: https://ruder.io/optimizing-gradient-descent/ (accessed on 29 December 2021).
26. Kingma, D.P.; Ba, J. Adam: A Method for Stochastic Optimization. In Proceedings of the 3rd International Conference for Learning Representations, San Diego, CA, USA, 7–9 May 2015. [CrossRef]
27. Fávero, L.P.; Belfiore, P.; de Freitas Souza, R. Chapter 22—Artificial neural networks. In *Data Science, Analytics and Machine Learning with R*; Academic Press: Cambridge, MA, USA, 2023; pp. 441–467. [CrossRef]
28. Dassault Systems Solidworks. Available online: https://www.solidworks.com/ (accessed on 15 April 2023).

machines

MDPI

Article

Effect of Hardening Temperature on Maraging Steel Samples Prepared by Direct Metal Laser Sintering Process

Radoslav Vandzura *, Vladimir Simkulet, Matus Gelatko, Michal Hatala and Zuzana Mitalova

Faculty of Manufacturing Technologies, Technical University of Košice, 080 01 Prešov, Slovakia
* Correspondence: radoslav.vandzura@tuke.sk; Tel.: +421-55-602-6359

Abstract: This paper deals with the application of the direct metal laser sintering (DMLS) process, which already has a dominant position in the area of additive manufacturing (AM). This DMLS technology is used in many branches of industry and medicine, especially in piece production, small series, and prototypes. The portfolio of used metal powder materials includes aluminum alloys, austenitic steels, maraging steels, special alloys of nickel and titanium. The properties of these products are very often improved by further heat treatment after printing, such as a hardening process, by which microstructure and hardness can be increased. Heat treatment processes of metal AM components are already described, but experiments focused on optimization of these processes are still missing. In the article, the maraging steel samples printed by the DMLS method are subjected to testing after hardening processes, which differ by reducing the maintaining time at a defined temperature, recommended by the manufacturer. The result of the evaluation will be the reaching of similar results, which are set by the powder manufacturer, however, with shorter time of samples treatment. Therefore, the elevated temperature is selected, with the purpose of monitoring the shortest possible time of a temperature impact. The experimental temperature was set 590 °C with different durations at this temperature, for 1, 2, 3, 4, 5 and 6 h. The cooling process runs controlled in the furnace or in the still air. The maintaining time proved to be the most ideal already at 1 h exposure and cooled in the still air, where a higher hardness value of around 50 HRC was reached. During the resulting microstructure evaluations, fine carbids and martensitic lamellae were observed. More uniform and finer lamellar microstructure occurred at 5 and 6 h temperature intervals.

Keywords: maraging steel; printing; laser; sintering; hardening; hardness; microstructure

1. Introduction

The direct metal laser sintering (DMLS) 3D printing method and other comparable printing techniques, such as Selective Laser Sintering (SLS) and Laser Powder Bed Fusion (LPBF), have already found use in a variety of technical fields, as well as in the medical field, aerospace industry, and other fields [1–3]. In addition to polymers, other materials can be used, including iron, non-ferrous metals, ceramics, and composites. [4–10]. By adding additive elements to the alloy [4–8], which is regularly experimented with in addition to finishing operations like heat treatment and hardening [2,5,7–11], the qualities of the material are improved. In the field of fast prototyping, the DMLS and SLS processes have their origins in the 1970s with the creation of prototypes. [1–6,8–11].

Due to the benefits they provide, particularly the more effective fabrication of components with extremely complicated geometries, interest in these technologies of additive manufacturing of metal parts has increased significantly over the past several years. Over the past ten years, a small number of businesses have begun to create DMLS technique variants with the intention of advancing process technology in general [12]. Figure 1 illustrates the evolution of DMLS through the year 2004 together with other related technologies by showcasing innovations and commercial developments in chronological order on the vertical axis and technological area on the horizontal axis.

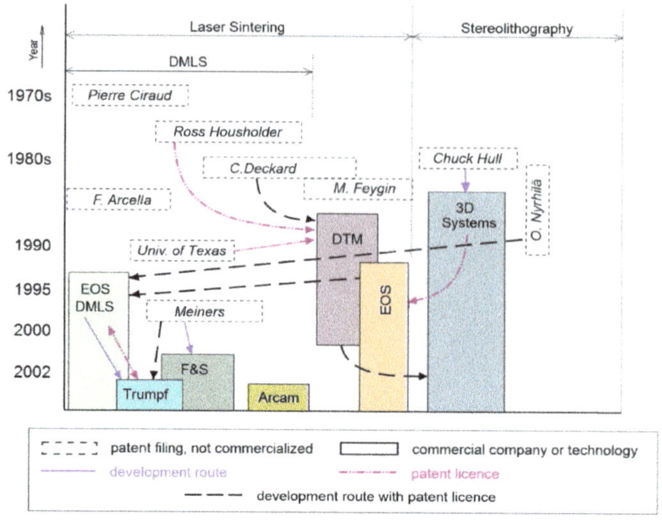

Figure 1. DMLS development and patent history since 1970.

A decade before additive manufacturing techniques and the name Rapid Prototyping became popularized in the 1960s of the 20th century, the notion of fabricating metal components via the laser-induced fusing of powder material was first proposed by Frenchman Pierre Ciraud [12,13]. His fundamental concept is to enable the production of parts without the need for casting molds. This idea was not sufficiently ready for commercialization, because computers were then only in their infancy. Six years after Ciraud, another inventor named Ross Householder submitted his patent application. He described a system and technique that were similar to subsequent commercial laser sintering systems [12,14].

The invention's claimed goal was to develop a novel molding method for building up three-dimensional components, using layers that could be controlled by computer technology. In one design, the layers are created using fusible particles, which are then individually fused by a laser beam to define the final result. Housholder was only able to thoroughly test a variant approach that did not require a laser at the time due to the prohibitively expensive cost of lasers at the time. His creation was not commercialized at the time and went unnoticed until DTM Corporation came upon it, while working on their own patent filings. [12,15].

Early through the 1980s, the initial steps in the commercialization of additive manufacturing techniques and powder-based additive processes were taken. Chuck Hull, the creator of 3D Systems and the US 4,575,330 patent [16], launched commercial "RP" in August 1984. In the area of laser sintering, EOS acquired exclusive rights to the full 3D system patent portfolio in 1997. Around 1986, Hull's approach was the subject of research by master's candidate Carl Deckard at the University of Texas (UT), who used powdered materials instead of liquid ones. Part Generation by Layerwise Selective Sintering (PGLSS) was the original name of his technique; Selective Laser Sintering was eventually used (SLS). A computer-aided laser apparatus that sequentially sinters a number of powder layers to create a desired portion layer-by-layer, was the subject of the ensuing October 1986 patent application [12,17].

In April 1987, Helisys was founded by Michael Feygin to commercialize his ideas, where he used a scanning laser in the production of laminated molds. Another similar development to cast shapes without the use of a mold took place by Frank Arcello, in March 1988. As a consequence of research and development by Deckard, Beaman, and associates at UT, DTM Corp. launched Sinterstation 2000 as the first commercially available system for laser sintering in December 1992. By EOS GmbH of Munich, Germany, the second industrial laser sintering system was introduced in April 1994. The combination of powder metallurgy technology from Electrolux Rapid Development (ERD) by Mr. Russia from Finland, and EOS plastic laser sintering technology led to the creation of the first commercial DMLS system.

A revolutionary powder idea for pressureless sintering with extremely little shrinkage was created by Nyrhilä in 1989 [18]. Before successfully reaching an agreement on a patent license and collaboration between ERD and EOS in 1994, he had the notion of employing this technique for DMLS and had discussed it with a number of possible partners. In 1994, the installation of the first test systems created by EOS [12], where the modified bronze nickel-based powder was sintered in 100 µm layers using a 100W CO2 laser, was done and commercial EO-SINT M 250 systems were deployed for another year. This opened the door for the first really commercial application of DMLS for quick machining and made it feasible to make complicated components with high accuracy and acceptable surface quality, which was previously impossible with any other direct metal technique. Since then, despite several changes in ERD's owners, the relationship remained. A German cooperative initiative was launched at the end of 1995 to improve the DMLS process further and make it possible to produce fully dense components by fully melting single component ceramic and metallic materials like 1.4404 steel [12].

Within the German cooperative project, at the end of 1995, together with partners EOS, Fockele and Schwarze (F&S) and Fraunhofer Institutes, developed equipment for the construction of fully dense parts from one-component ceramic and metal materials, using DMLS technology.

Due to the small market size and prospects for machinery sales, the Trumpf company terminated PBF between 2006 and 2014 [13]. In 2002, f. Arcam invented the technique for electron beam powder bed fusion and unveiled the first industrial system. In parallel with Arcam, the Extrude Hone Company unveiled the first AM system based on metal sintering in 1999 [15].

Systems from traditional machine makers, such as Trumpf and DMG, followed suit during the following two decades. (Figure 2.) The first business to adopt a traditional plasma arc welding procedure for AM machine technology was Norsk Titanium in 2007. The aerospace industry was the major area of attention. HP and Desktop Metal introduced their metal printer technologies in 2016 [15], entering the market.

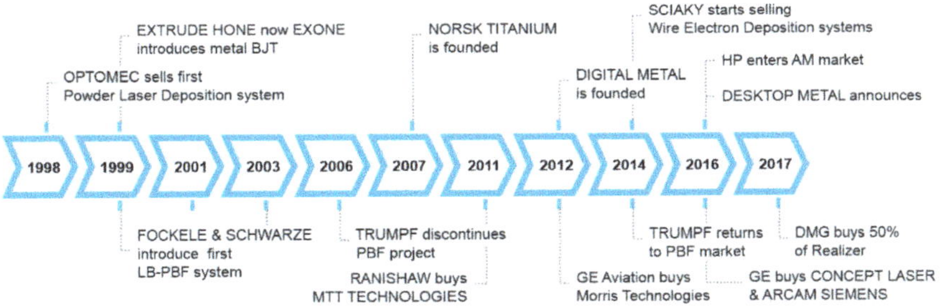

Figure 2. Summarized development of DMLS history after 2004 [15].

Other operations, including as the curing process, enhance the final qualities of the objects produced by these printers. Some researchers [19–21] studied the mechanical characteristics of powder additive products and noted alterations in the material's properties or microstructures after the printing process or after the operation was finished [22,23].

According to several publications [7–9,24–29], these qualities have recently been enhanced by heat treatment procedures (such as tempering, annealing, hardening, etc.), where the values of hardness and strength have more than doubled. Another of the researched features, that is a part of other essential usable properties of the product generated by this method of 3D printing technology, is the fatigue strength of the material, prepared by DMLS or a similar employed laser melting technique [30,31].

The heat treatment, annealing for example, demonstrably changes mechanical properties of AM materials in the study: Production of Hybrid Joints by Selective Laser Melting of Maraging Tool Steel 1.2709 on Conventionally Produced Parts of the Same Steel, by K. Kucerová, et. al, where hybrid parts were made by DMLS technology, from maraging steels 18Ni300 (MS1) and VACO180. A study deals with various ways of annealing and cooling of mentioned hybrid material, and the influence of these process on mechanical properties of the material, is taken into account. Original tensile strength was 1029 MPa in components. Using the solution annealing, yield strength was decreased by 150 MPa. In the case of precipitation annealing, a strength increased to 2011 MPa [24].

During the experiment by Luca Fachini, et. al: Ductility of a Ti-6Al-4V alloy produced by selective laser melting of pre-alloyed powders, martensitic structure was stabilized, due to the process variation, which consequently led to the increasing of mechanical properties of Ti-6Al-4V alloy [7]. Some other researches are focused on the resulting mechanical properties of the material, reached by the input AM process parameters variation, however, not by the material properties changing after the 3D printing process. There are not too many available sources and scientific studies, which are aimed on the MS1 material properties variation, or any similar maraging steels, prepared by the AM process. Experiments on conventionally made maraging steel materials, could be the good stepping-stones, such as studies [32,33]. Nevertheless, these materials includes apparently the same chemical compositions, a different manufacturing method need to be taken into account, which refers to AM materials prepared using the PBF technologies. Available AM materials generally have one heat treatment process, in order to increase properties of individual material. However, it is appropriate to pay attention on researches, focused on various heat treatments, with the aim to recognize the behavior of material, its mechanical properties and inner structure changes at different ways and conditions of heat treatment.

Evaluated material reaches double values of hardness and strength increase, after the heat treatment with 490 °C temperature for 5 h. Examined lower temperature did not reach such an effect [20,21,24,28]. Therefore, the plan of the experiment will be the reaching of similar results and evaluate the best possible time of elevated temperature impact on the sample, using the epistemological process.

2. Materials and Methods

Direct metal laser sintering (DMLS) means sintering using a laser beam, where metal powder is directly used to produce metal connections in the construction process. A diverse portfolio of materials used in the production of metal parts by DMLS, developed by EOS GmbH, as the first commercial method of rapid prototyping [1–3,6,8], includes Tool Steels, Stainless Steels and Maraging Steels, various Nickel Alloys, Titanium, Refractory Metals, light metals as Aluminum, Copper etc. [1–5,8].

Direct metal laser sintering (DMLS), electron beam melting (EBM), selective heat sintering (SHS), selective laser melting (SLM) and selective laser sintering (SLS) are some of the most popular printing methods utilized in the Powder Bed Fusion process (PBF), according to ISO /ASTM 52900:2021 standard. Laser or electron beams are used in powder bed fusion (PBF) techniques to melt and fuse separate material powder particles. The powder material must be dispersed over earlier layers in every PBF procedure. This can be accomplished using a variety of devices, such as a roller or a blade. The supply of new materials is provided by a hopper or reservoir beside the bed. The same process as selective laser sintering (SLS), direct metal laser sintering (DMLS) uses metals rather than polymers. Layer-by-layer, the powder is sintered during the process. The basic principle of PBF technology is shown in Figure 3 [34–36].

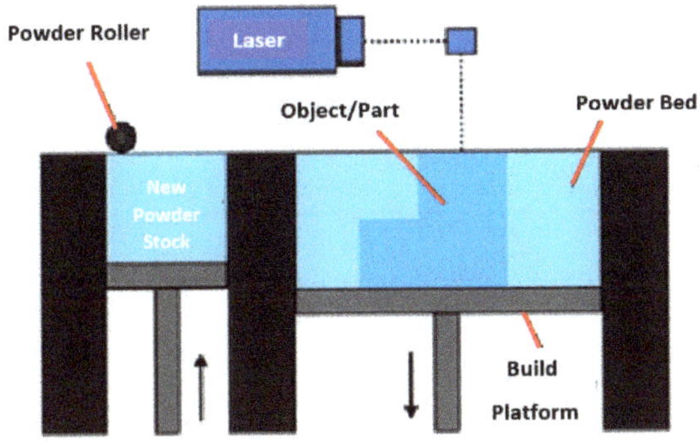

Figure 3. Principle of PBF and other similar technologies [36].

Advantage of DMLS process in comparison other similar methods (such as SLS, LPBF and more), that it provides higher resolution of a details, thanks to the use of thinner layers, which allows a smaller diameter of powder particles. This ability makes it possible to create more complex parts of shapes [3,4], which are also used in the field of comfort cooling and small structural components with specific shapes. Heat treatment of these printed products could optimize and improve their properties [4,6–8,13]. The manufacturer recommends a prescribed temperature of 490 °C for 6 h during the curing process.

The experiment is based on a heat treatment (age hardening) of MS1 material samples, for 1–6 h at the increased hardening temperature equal to 590 °C. The main aim of the experiment is to evaluate the variation of hardness and microstructure within the MS1 material at the various durations of hardening process at a certain temperature. Concretely, at a 590 °C for 1, 2, 3, 4, 5 and 6 h. Additionally, final structure and hardness were evaluated after the cooling influence, as for the cooling in the still air, so for the cooling in the furnace. Heat treatment conditions for different material samples are summarized in Table 1.

Table 1. Conditions of experimental heat treatment process.

Sample	Heat Treatment		
	Duration Time (Hours)	Curing Temperature (°C)	Cooling
1.0	1	590	in the still air
2.0	2	590	in the still air
2.1	2	590	in furnace
3.0	3	590	in the still air
4.0	4	590	in the still air
5.0	5	590	in the still air
6.0	6	590	in the still air
6.1	6	590	in furnace

– furnace temperature is raised to curing temperature in 30 min.

For investigated experiment, maraging steel made by the DMLS production method, was chosen, marked by the manufacturing company as MS1. This powder material is optimized for the application in EOSINT company 3D printing systems.

A metal powder of this maraging steel with a particle diameter from 20 to 80 μm was used, which was bonded without the need of a binder, using a high-power laser beam [1–8].

EOS Maraging Steel MS1 powder material corresponds to the American classification: 18% Ni Maraging 300, EU standard norm 1.2709, and German norm classification X3NiCoMoTi 18-9-5 [37]. This material can be easily heat-treated, using the heat hardening process, which provides excellent hardness and strength [38]. The chemical composition of tested material [37–39] is shown in Table 2.

Table 2. Chemical composition of investigated material.

Element *	Al	Ti	Mn, Si	Cr, Cu	Mo	Co	Ni
Mass (%)	0.05 – 0.15	0.6 – 0.8	up to 0.1	up to 0.5	4.5 – 5.2	8.5 – 9.5	17 – 19

* C is below 0.03 percent, P and S is below 0.01 percent, Fe makes up the rest.

Components or semi-components printed in this way manufactured of maraging steel on the printer from EOS are easy to machine after the build process and can be very easy to improve after finishing operations such as hardening. In their research papers, various authors describe the improvement of properties by curing at 490 °C for 6 h [20,28,29], where they reached an almost twice as high strength and hardness value. For our experiment, we will use a higher temperature and a shorter dwell time on the component to achieve a similar effect. The mechanical properties reporting by manufacturer [38,39], are listed in Tables 3 and 4.

Table 3. Mechanical Properties as built (a) and printed and also hardened (b).

Density ρ, (g/cm^3)		Tensile Strength (MPa)	Yield Strength Rp0.2, (%)	Elongation at Break, %	Modulus of Elasticity, (GPa)	Hardness (HRC) *
direction 8.0–8.1 direction	(a)	XY 1000–1200 Z	XY 1950–1150 Z	XY 6–14 Z	XY 135–185 Z	33–37
8.0–8.1	(a)	1000–1200	1850–1150	-	130–170	
	(b)	1950–2150	1090–1290	2–6		50–56

* Hardness measured on polished surface.

Table 4. Thermal Properties of Parts recommended by the manufacturer.

	As Built	Hardened *
Thermal conductivity	14.2–15.8 W/m °C	19–21 W/m °C
Specific heat capacity	430–470 J/kg °C	
Operating temp.	max. 400 °C	

* Hardened temperature at 490 °C/ 6 h, cooling on the air.

2.1. Experiment Preparation

The samples were produced on additive manufacturing machine by the EOSINT company, with the designation EOSINT M280, intended for additive manufacturing of metal products. This 3D printer has 250 × 250 × 325 mm volume of chamber. This type employs the Yb fiber laser with a maximum 400 W beam power, up to 700 mms^{-1} scan speed, and 80 μm focus diameter. EOS Maraging Steel MS1 was used for the sample preparation. A shape and a size of the input powder material MS1 is plotted in Figure 4. A powder was observed using the scanning electron microscope SEM f. Tescan type Mira 3 and chemical composition was evaluated using the EDX analyser f. Oxford. A various size of spherically shaped particles can be observed. Experimental powders, evaluated using the spectral EDX analysis (Figure 4b), showed values corresponding to the values of the manufacturer, in a wider spectrum (Table 2).

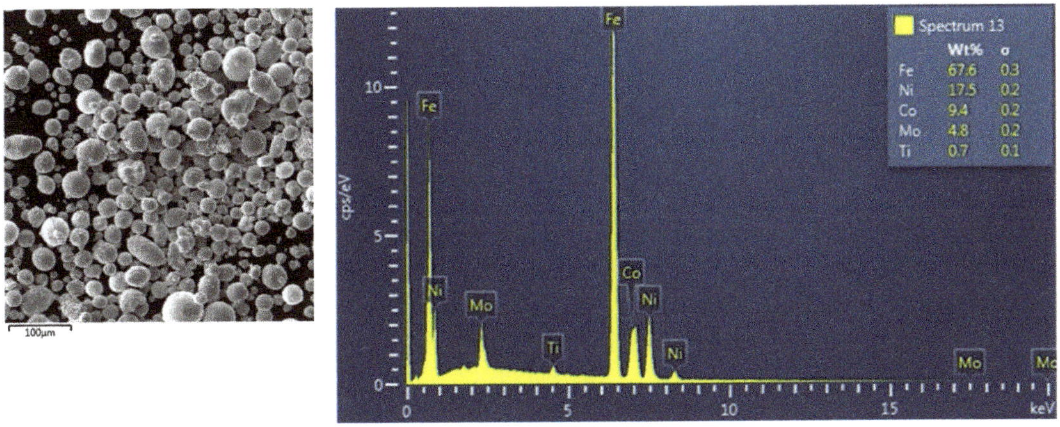

(a) **(b)**

Figure 4. A shape and a size of powder material MS1, evaluated using SEM Tescan Mira 3 (**a**), EDX analysis of powder, using f. Oxford analyzer (**b**).

EOSINT M280 AM system is able to print the component, based on the. stl file, straight from the CAD designs. For melting the powder, the laser power was adjusted to 200 W, the scanning speed to 650 mms^{-1} and the hatching distance to 80 μm. Layer thickness was set on 40 μm for the layering process, and throughout the PBF process, the whole printing system was shielded by argon (Ar) gas. Meander printing strategy, with 100% infill setting, was used. Dimensions of tested samples were (XxZxY) 10 × 10 × 30mm. Dimensions and a shape of the material during printing, are shown in Figure 5. The layering of the material, during the DMLS process, was in the direction of the Z axis.

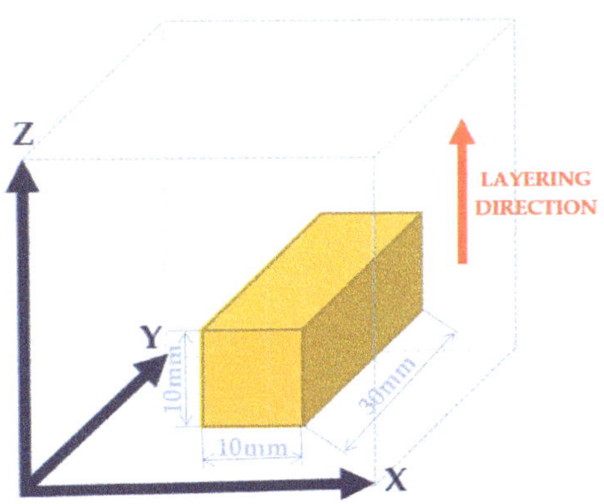

Figure 5. Scheme of the principle of DMLS printing for structure creation.

2.2. Process of Thermal Treatment

Heat treatment by aging (age hardening) took place in the LAC device (Figure 6). The LAC device is an electric induction furnace with the possibility of setting temperature profiles. The chosen temperature profile was from a room temperature of 20 °C to a curing temperature of 590 °C in 30 min and maintaining at a temperature of 590 °C for the entire experiment for 6 h. The raising to the curing temperature was monitored using a FLIR T1030SC thermal camera (Figure 7).

Figure 6. Induction furnace LAC.

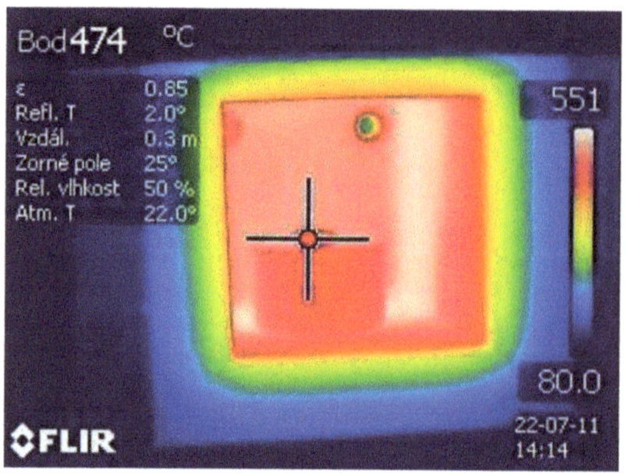

Figure 7. Verification of the raising temperature using a FLIR T1030SC thermal camera.

Prepared samples of the investigated material were gradually hardened in an induction furnace with an initial run-up time of 30 min. to a temperature of 590 °C. After the duration of 1, 2, 3, 4, 5 and 6 h, were taken out on a concrete pad and cooled in the still air, at a room temperature of 20 °C. To compare the effect of cooling in the air and cooling in the furnace, samples prepared at 2 and 6 h were cooled differently. One was cooled in the turned off furnace and the other was removed and cooled in the still air.

2.3. Preparation of Samples for Microstructure Observation and Hardness Evaluation

Selected places for the evaluation of hardness and microstructure were chosen in the center of the sample, in the cross section of the x-z axis and the x-y axis, in order to observe the effect of heat treatment inside the material. The observed locations are marked in Figure 8. Material samples after heat treatment were split by abrasive water jet (AWJ) technology. AWJ technology is suitable for cutting of samples, because it does not thermally affect the material [40]. Samples for microstructure observation were prepared by grinding on the f. MTH Hrazdil device with the help of water cooling, sandpapers with grits of 180,

320, 500, 1200, 2000 and 4000. Then, samples were polished on a satin cloth with applied diamond polishing paste 0.1 μm.

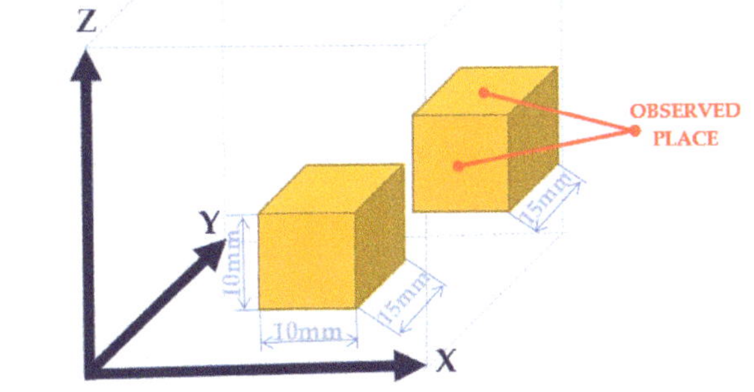

Figure 8. Section of material samples after temp. processing with a marked point of evaluation of hardness and microstructure.

The material samples were then observed with a Nikon MA100 light microscope and evaluated on Rockwell Insize ISH-RSR400 hardness testers. Cooling temperature was measured with thermal camera FlirT1030SC.

3. Results and Discussion

Measured cooling temperature dependences of MS1 material samples after heat treatment at a temperature of 590 °C, taken from the induction furnace after 1, 2, 3, 4 and 6 h, were cooled in the still air at a room temperature of 20 °C. Observation of the cooling process lasted approximately 20–30 min. Temperature was measured with thermal camera (Figure 9). Figure 10 shows the temperature dependence of the sample that was cooled in the furnace.

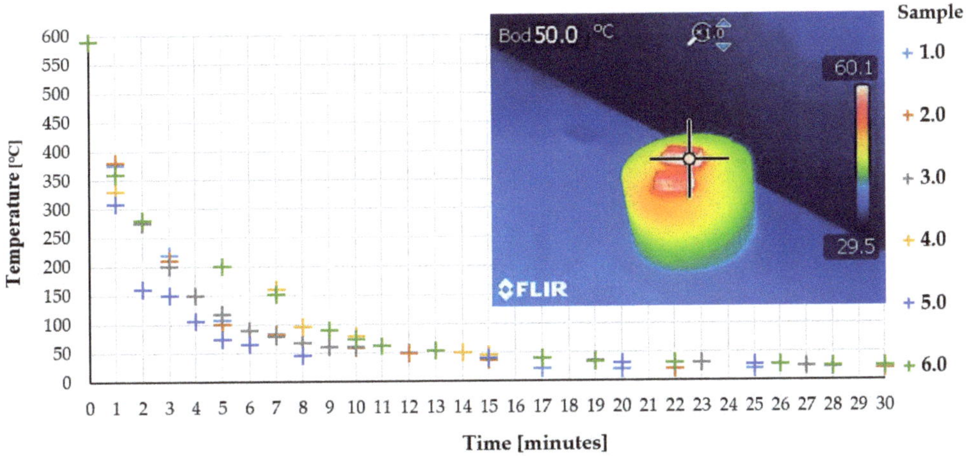

Figure 9. Temperature curve of cooling in the still air.

Tables below show Rockwell hardness measurements on samples with different lengths of used heat treatment process. Table 5 shows the hardness measurement at the center of the sample surface on each side. Table 6 shows the hardness measurement in the cross section of the sample. Sample 0.0 is shown for comparison, printed only, without heat treatment. The first sample number is the duration value, e.g., 1.0—length of maintaining at a temperature for 1 h.

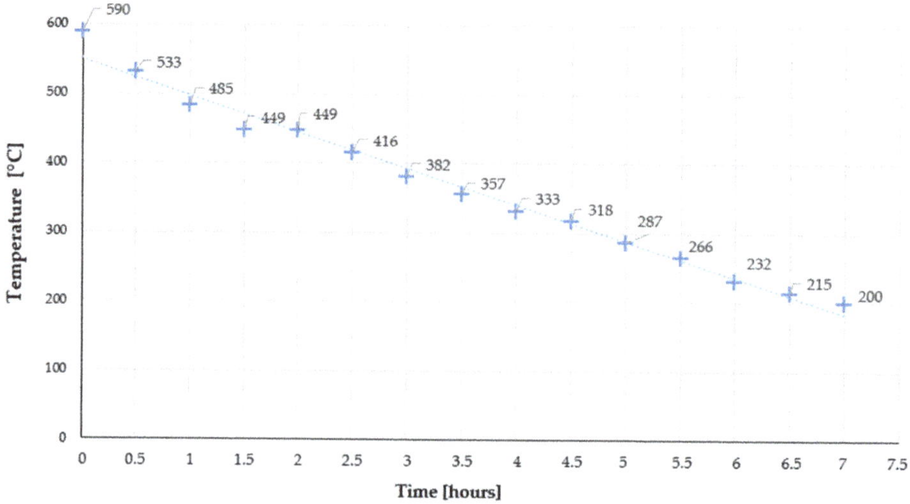

Figure 10. Temperature curve of cooling in the furnace.

Table 5. Rockwell hardness measurement values at the center of the sample surface.

Sample	Measuring Side, Hardness Value				Average Hardness Value
	x-y(Bottom)	x-y(Top)	y-z(Right Side)	y-z(Left Side)	
0.0 *		36.7			36.7
1.0	29.6	50.4	47.8	47.9	43.9
2.0	27.1	48.1	36.7	45.2	39.3
2.1 **	35.1	45.9	37.9	48.8	41.9
3.0	40.5	37.9	42.4	41.1	40.5
4.0	25.1	37.3	46.4	45.7	38.6
5.0	32.5	46.3	37.5	47	40.8
6.0	37	51.3	42.7	45.8	44.2
6.1 **	21.9	42.7	49.8	46.8	40.3

*—only printed, without heat treatment. **—after cooling in furnace.

Hardness values after different lengths of stay at the tested temperature of 590 °C in all cases compared to only printed samples showed a higher hardness. Even after 1 h of holding at the temperature, average hardness was 44 HRC and in the cut surface 51 HRC. The lower value for these samples was distorted by the measurement at the bottom of the sample, where can be assumed that the surface of the construction platform is still insufficiently sanded. This artifact appears in every sample where the hardness values are clearly lower. During the measurement in the cut surface, this phenomenon did not occur, as we performed the measurement about 1 mm below the surface. With a longer duration for 2, 3, 4, 5 and 6 h, the hardness values decreased slightly. In the experimental evaluation of cooling in the still air—marked samples 2.0 and 6.0 and in the furnace—samples marked 2.1 and 6.1, these showed a slight increase in hardness only on the surface. The center of the sample did not change or slightly decreased. Therefore, can be assumed that the most ideal time for increasing the hardening temperature from 490 °C to 590 °C is 1 h and free cooling in the air.

Table 6. Rockwell hardness measurement values in the cross section of the sample.

Sample	Measuring Side, Hardness Value				Average Hardness Value	Measurement in the Center
	x-z (Bottom)	x-z (Top)	x-z (Right Side)	x-z (Left Side)		
1.0	49.7	50.2	51	50.7	50.4	51
2.0	48.9	48.2	49.1	43.4	47.4	49.2
2.1 **	49.4	49.9	50.2	49.5	49.8	49.5
3.0	49.2	46.5	48.6	48.1	48.1	49.5
4.0	48	45.2	47.7	46.9	47	47.7
5.0	44.1	46	45.9	45.4	45.4	46.9
6.0	46.8	47.5	46.8	45.4	46.6	47.5
6.1 **	47	46.7	46	46	46.4	46.8

**—after cooling in furnace.

The following Figure 11 shows observed microstructures of samples before heat treatment in the x-z and x-y planes The microstructure in the y-z plane was similar to that in the x-z plane as these are cross-sectional planes. In the x-y plane, the trace after the laser melting of the particles is visibly larger compared to x-z and y-z. This is due to the longer trace of the laser beam movement. At both X-Z and X-Y planes, features characteristic for DMLS 3D printing technology were observed. In the evaluated planes, direction and shape of the laser layering, were observed. Additionally, individual layers during sintering of the material at the uniform distance of about 0.1 μm, equal to the laser displacement, can be observed.

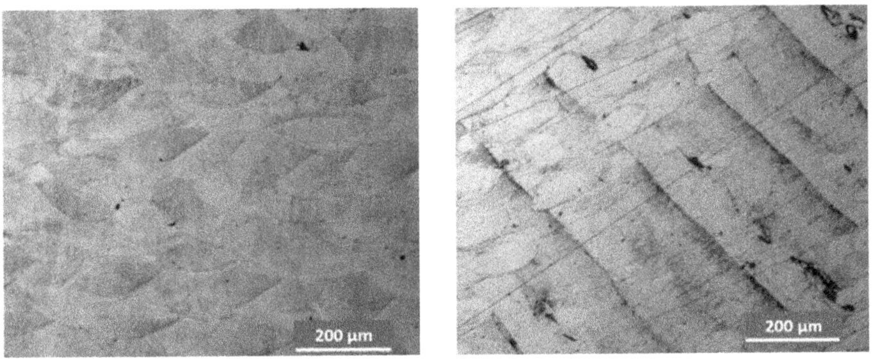

(a) (b)

Figure 11. Microstructure of sample without hardening process, only printed (sample 0.0), x-z plane (**a**), x-y plane (**b**), 100× magnification.

Observed microstructures of the MS1 material after heat treatment (Figures 12–14), showed similar layered formations typical of 3D printed samples. The microstructure of all heat-treated samples was uniform throughout the sample cross section. At a higher magnification of 1000x, a similar detail of the remelted particle with fine carbides can be seen again, with places of martensite lamellae. The microstructure at 1 and 2 h showed fine lamellae after the heat treatment, but not uniformly distributed, compared to the longer time of the heat process. At the 5 and 6 h interval, microstructures were quite similar and evenly distributed. In the 6 h interval of heat treatment, these lamellae are significantly softer, which is also indicated by the resulting hardness, which no longer increased rapidly. On the contrary, it was lower.

(**a**) (**b**)

Figure 12. Microstructure of the sample 1 h stay at 590 °C (sample 1.0) 1000× magnification (**a**); sample with length 2 h stay at 590 °C (sample 2.0) 1000× magnification (**b**).

(**a**) (**b**)

Figure 13. Microstructure of the sample 3 h stay at 590 °C (sample 3.0) 1000× magnification (**a**); sample with length 4 h stay at 590 °C (sample 4.0) 1000× magnification (**b**).

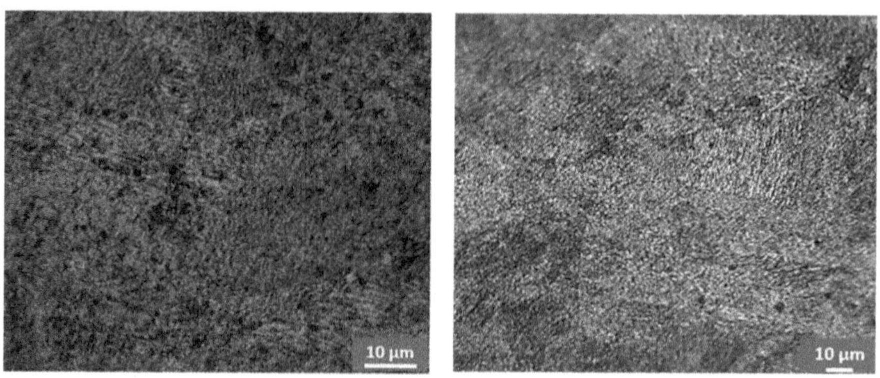

(**a**) (**b**)

Figure 14. Microstructure of the sample 5 h stay at 590 °C (sample 5.0) 1000× magnification (**a**); sample with length 6 h stay at 590 °C (sample 6.0) 500× magnification (**b**).

4. Conclusions

The article summarized the following claims:

Used heat treatment had an effect on pointing to a change in the length of stay at a temperature of 590 °C. The maintaining time proved to be the most ideal already at 1 h exposure and cooled in the still air, where a higher hardness value of around 50 HRC was reached. A longer maintaining time at 590 °C had the same or a slight decrease in hardness values. The manufacturer indicates a residence time for 6 h, at a temperature of 490 °C, where the values are around the level of 50 HRC. The microstructural evaluation was typical for the heat treatment, where fine carbides and martensitic lamellae were observed with a slight increase in hardness values in the microstructure. A more uniform and finer lamellar microstructure was shown with a longer interval of temperature exposure at 5 and 6 h intervals. This also indicates that the resulting hardness, which no longer increased rapidly, on the contrary, was somewhat lower, indicating a more relaxed state of the material after the hardening process.

Author Contributions: Conceptualization, V.S. and R.V.; investigation, R.V. and V.S.; data curation, R.V. and M.G.; original writing draft preparation, R.V. and V.S.; review writing and editing, V.S., R.V. and M.G.; visualization, Z.M.; supervision, V.S. and M.H. All authors have read and agreed to the published version of the manuscript.

Funding: This research was funded by the grant projects: KEGA no. 018TUKE-4/2021, VEGA 1/0391/22, APVV-21-0228 and ITMS no. 313011BWN5.

Data Availability Statement: Not applicable.

Acknowledgments: This work was funded by the Slovak Research and Development Agency under contract No APVV-21-0228 and the projects VEGA1/0391/22, KEGA 018TUKE-4/2021 were granted by the Ministry of Education, Science, Research and Sport of the Slovak Republic and this publication is the result of the Project implementation: Development of excellent research capacities in the field of additive technologies for the Industry of the 21st century, ITMS: 313011BWN5, supported by the Operational Program Integrated Infrastructure funded by the ERDF.

Conflicts of Interest: The authors declare no conflict of interest. The founding sponsors had no role in study design; when collecting, analyzing, or interpreting data; while writing the manuscript; or in the decision on the publication of the results.

References

1. Mançanares, C.G.; de S. Zancul, E.; Cavalcante da Silva, J.; Miguel, P.A.C. Additive manufacturing process selection based on parts selection criteria. *Int. J. Adv. Manuf. Technol.* **2015**, *80*, 1007–1014. [CrossRef]
2. Yasa, E.; Kruth, J.-P. Microstructural investigation of selective laser melting 316L stainless steel parts exposed to laser re-melting. *Procedia Eng.* **2011**, *19*, 389–395. [CrossRef]
3. Palermo, E. Laserové spekanie. Available online: https://www.livescience.com/38862-selective-laser-sintering.html (accessed on 8 August 2019).
4. Dobránsky, J.; Kočiško, M.; Baron, P.; Simkulet, V.; Běhálek, L.; Vojnová, E.; Nováková-Marcinčínová, L'. Evaluation of the Impact Energy of the Samples Produced by the Additive Manufacturing Technology. *Metalurgija* **2016**, *55*, 477–480.
5. Monková, K. *Some Aspects of DMLS Technology*; RISE Association: Houston, TX, USA, 2017; p. 117. ISBN 978-80-87670-23-1.
6. Mower, T.M.; Long, M.J. Mechanical behavior of additive manufactured, powder-bed laser-fused materials. *Mat. Sci. Eng. A* **2016**, *651*, 198–213. [CrossRef]
7. Brandl, E.; Leyens, C.; Palm, F.; Schoberth, A.; Onteniente, P. Wire instead of powder? Properties of additive manufactured Ti–6Al–4V for aerospace applications. In Proceedings of the Euro-uRapid International Conference on Rapid Prototyping, Rapid Tooling & Rapid Manufacturing, Berlin, Germany, 23–24 September 2008.
8. Facchini, L.; Magalini, E.; Robotti, P.; Molinari, A.; Höges, S.; Wissenbach, K. Ductility of a Ti–6Al–4V alloy produced by selective laser melting of pre-alloyed powders. *Rapid Prototyp. J.* **2010**, *16*, 450–459. [CrossRef]
9. Van Hooreweder, B.; Boonen, R.; Moens, D.; Kruth, J.-P.; Sas, P. On the determination of fatigue properties of Ti6Al4V produced by selective laser melting. In Proceedings of the 53rd AIAA/ASME Structures, Structural Dynamics and Materials Conference, Honolulu, HI, USA, 23–26 April 2012.
10. Liu, Q.; Elambasseril, J.; Sun, S.; Leary, M.; Brandt, M.; Sharp, P.K. The effect of manufacturing defects on the fatigue behavior of Ti–6Al–4V specimens fabricated using selective laser melting. *Adv. Mater. Res.* **2014**, *891–892*, 1519–1524. [CrossRef]

11. Radchenko, S.; Knapčíková, L.; Dupláková, D.; Hatala, M. Verifikácia mechanických vlastností kompozitných materiálov v klimakomore. *Strojárstvo/Strojírenství. Roč.* **2016**, *10*, 82. (accessed on 1 October 2016).
12. Shellabear, M.; Nyrhilä, O. DMLS—Development History and State of the Art. Available online: https://www.i3dmfg.com/wp-content/uploads/2015/07/History-of-DMLS.pdf (accessed on 15 December 2022).
13. Ciraud, P. Verfahren und Vorrichtung zur Herstellung beliebiger Gegenstände aus beliebigem schmelzbarem Material. German Patent Application DE 2263777, 28 December 1971.
14. Housholder, R. Molding Process. US Patent 4,247,508, 3 December 1979.
15. Metal Additive Manufacturing History. Available online: https://additive-manufacturing-report.com/technology/metal/metal-additive-manufacturing-history/ (accessed on 15 November 2019).
16. Hull, C. Method and Apparatus for Production of Three-Dimensional Objects by Stereolithography. US Patent 4,575,330, 8 August 1984.
17. Deckard, C. Method and Apparatus for Producing Parts by Selective Sintering. US Patent 4,863,538, 17 October 1986.
18. Nyrhilä, O.; Syrjälä, S. Manufacture of dimensionally precise pieces by sintering. Finnish Patent Finnish Patent 91725, 7 April 1989.
19. Gašpár, Š.; Paško, J. Analysis of influence of pressing speed, of melt temperature and of casting position in a mold upon ultimate tensile strength Rm of die casting from aluminium. In *Advanced Materials Research: 2nd International Conference on Manufacturing 2014, Singapore, 9–10 February 2014*; Springer: Berlin/Heidelberg, Germany, 2014; Volume 909, pp. 3–7. ISBN 978-303835059-0. ISSN 1022-6680.
20. Simkulet, V.; Gusak, O.; Zajac, J.; Parilák, L'.; Mitaľová, Z. Verifikácia procesu tepelného spracovania materiálu vyrobeného technológiou priameho spekania laserom. In *Automatizácia a riadenie v teórii a praxi: ARTEP 2018*; Technical University of Košice: Košice, Slovakia, 2018; pp. 1–7. ISBN 978-80-553-2914-7.
21. Monkova, K.; Monka, P.; Mandulak, D. Hardness and Surface Roughness Study of Steel Part Produced by Additive Technology. *DEStech Trans. Eng. Technol. Res* **2017**. [CrossRef] [PubMed]
22. Yu, N. Process Parameter Optimization for Direct Metal Laser Sintering (DMLS). Ph.D. Thesis, National University of Singapore, Singapore, 2005.
23. Maharjan, N.; Wu, N.; Zhou, W. Hardening Efficiency and Microstructural Changes during Laser Surface Hardening of 50CrMo4 Steel. *Metals* **2021**, *11*, 2015. [CrossRef]
24. Kučerová, L.; Zetková, I.; Jeníček, Š.; Burdová, K. Production of Hybrid Joints by Selective Laser Melting of Maraging Tool Steel 1.2709 on Conventionally Produced Parts of the Same Steel. *Materials* **2021**, *14*, 2105. [CrossRef] [PubMed]
25. Sarafan, S.; Wanjara, P.; Gholipour, J.; Bernier, F.; Osman, M.; Sikan, F.; Molavi-Zarandi, M.; Soost, J.; Brochu, M. Evaluation of Maraging Steel Produced Using Hybrid Additive/Subtractive Manufacturing. *J. Manuf. Mater. Process.* **2021**, *5*, 107. [CrossRef]
26. Wang, X.; Chou, K. Effect of support structures on Ti-6Al-4V overhang parts fabricated by powder bed fusion electron beam additivemanufacturing. *J. Mater. Process. Technol.* **2018**, *257*, 65–78. [CrossRef]
27. Gåård, A.; Krakhmalev, P.; Bergström, J. Microstructural characterization and wear behavior of (Fe,Ni)–TiC MMC prepared by DMLS. *J. Alloys Compd.* **2006**, *421*, 166–171. [CrossRef]
28. Živčák, J.; Nováková-Marcinčínová, E.; Nováková-Marcinčínová, L'.; Balint, T.; Puškár, M. Increasing Mechanical Properties of 3D Printed Samples by Direct Metal Laser Sintering Using Heat Treatment Process. *J. Mar. Sci. Eng.* **2021**, *9*, 821. [CrossRef]
29. Zhao, Z.-Y.; Li, L.; Bai, P.-K.; Jin, Y.; Wu, L.-Y.; Li, J.; Guan, R.-G.; Qu, H.-Q. The Heat Treatment Influence on the Microstructure and Hardness of TC4 Titanium Alloy Manufactured via Selective Laser Melting. *Materials* **2018**, *11*, 1318. [CrossRef] [PubMed]
30. Avanzini, A.; Battini, D.; Gelfi, M.; Girelli, L.; Petrogalli, C.; Pola, A.; Tocci, M. Investigation on fatigue strength of sand-blasted DMLS-AlSi10Mg alloy. *Procedia Struct. Integr.* **2019**, *18*, 119–128. [CrossRef]
31. Rafi, H.K.; Starr, T.L.; Stucker, B.E. A comparison of the tensile, fatigue and fracture behavior of Ti–6Al–4V and 15–5 PH stainless steel parts made by selective laser melting. *Int. J. Manuf. Technol.* **2013**, *69*, 1299–1309. [CrossRef]
32. Simm, T.H.; Sun, L.; Galvin, D.R.; Hill, P.; Rawson, M.; Birosca, S.; Gilbert, E.P.; Bhadeshia, H.; Perkins, K. The effect of a two-stage heat-treatment on the microstructural and mechanical properties of a maraging steel. *Materials* **2017**, *10*, 1346. [CrossRef] [PubMed]
33. Shamantha, C.R.; Narayanan, R.; Iyer, K.J.L.; Radhakrishnan, V.M.; Seshadri, S.K.; Sundararajan, S.; Sundaresan, S. Microstructural changes during welding and subsequent heat treatment of 18Ni (250-grade) maraging steel. *Mater. Sci. Eng. A* **2000**, *287*, 43–51. [CrossRef]
34. N'Dri, N.; Mindt, H.W.; Shula, B.; Megahed, M.; Peralta, A.; Kantzos, P.; Neumann, J. DMLS Process Modeling and Validation. In Proceedings of the TMS 2015 144th Annual Meeting & Exhibition, Orlando, FL, USA, 15–19 March 2015; pp. 389–396.
35. Sun, S.; Brandt, M.; Easton, M. Powder bed fusion processes: An overview. In *Laser Additive Manufacturing: Materials, Design, Technologies, and Applications*; Woodhead Publishing Series in Electronic and Optical Materials; Elsevier: Amsterdam, The Netherlands, 2017; pp. 55–77.
36. About Additive Manufacturing. Available online: https://www.lboro.ac.uk/research/amrg/about/the7categoriesofadditivemanufacturing/powderbedfusion/ (accessed on 22 December 2022).
37. Edwards, P.; Ramulu, M. Fatigue performance of selective laser melted Ti–6Al–4V. *Mater. Sci. Eng. A* **2014**, *598*, 327–337. [CrossRef]
38. Available online: https://www.scribd.com/document/479137808/EOS-MaragingSteel-MS1-en-pdf# (accessed on 15 September 2022).

39. Available online: https://www.advancedtek.com/wp-content/uploads/2021/09/EOS_materials_table_overview_M_en_WEB.pdf (accessed on 15 December 2022).
40. Lebar, A.; Junkar, M. Simulation of abrasive waterjet machining based on unit event features. *Proc. Inst. Mech. Eng. Part B J. Eng. Manuf.* **2003**, *217*, 699–703. [CrossRef]

MDPI AG
Grosspeteranlage 5
4052 Basel
Switzerland
Tel.: +41 61 683 77 34

Machines Editorial Office
E-mail: machines@mdpi.com
www.mdpi.com/journal/machines

www.ingramcontent.com/pod-product-compliance
Lightning Source LLC
Chambersburg PA
CBHW061622210926
51850CB00009B/2405